D0935355

Physical Approach to Short-Term Wind Power Prediction

Matthias Lange Ulrich Focken

Physical Approach to Short-Term Wind Power Prediction

With 89 Figures and 13 Tables

Dr. Matthias Lange
Dr. Ulrich Focken

Energy & meteo systems GmbH
Marie-Curie-Str. 1
26129 Oldenburg
Germany
E-mail: matthias.lange@energymeteo.de
ulrich.focken@energymeteo.de

Library of Congress Control Number: 2005931222

ISBN-10 3-540-25662-8 Springer Berlin Heidelberg New York
ISBN-13 978-3-540-25662-5 Springer Berlin Heidelberg New York

Springer is a part of Springer Science+Business Media
springeronline.com

Printed in The Netherlands

Typesetting: by the authors and TechBooks using a Springer LaTeX macro package
Cover design: *design & production* GmbH, Heidelberg

Printed on acid-free paper SPIN: 11419068 55/TechBooks 5 4 3 2 1 0

Foreword

Triggered by a discussion on the nature of future electricity supplies, wind energy utilisation has boomed dramatically, first in the United States of America and Denmark and later in Germany and Spain. Thanks to state subsidies, it has within 15 years overtaken the volume of the classic renewable hydro-power, and today it accounts for about 5% of electricity generation.

Two factors set off this development: an awareness of the limited availability of fossil fuels and the recognition that in the 19th and 20th centuries the massive release of fossil CO_2 had kicked off a gigantic climate experiment whose results remain unpredictable. The discussion on the side effects of the wind energy boom, such as occupation of land and the challenges presented by integration into conventional electricity generation systems, frequently distract attention from the real goals and benefits of this technology. These are establishing an energy sector that will, in the short term, reduce CO_2 emissions and the exploitation of finite resources and, in the long term, create an unlimited sustainable energy supply.

Because fossil reserves are relatively easy to exploit, a system developed that could hardly be more convenient. It makes electric power available in large quantities at moderate prices and in a way that is easy to plan. The task of the power utility is essentially limited to "uncritically" adjusting the supply from central power stations to the demand from consumers. A low-CO_2 sustainable energy sector demands different standards. Wind and solar power have a high potential, but they are subject to high natural fluctuations and, in general, are connected to the electricity grid in a decentral way. The share of future storage technologies such as hydrogen technology will be as small as possible for reasons of efficiency and cost.

So, future-compatible electricity generation will comprise many different, partly innovative components, which also demands a considerable research and development effort. On the one hand, there is the fluctuating input from renewable sources, and on the other, electricity consumption that must to a certain extent be adjusted to supply by means of intelligent solutions. In between, to a declining extent, modern conventional energy producers, such as coal and natural gas power stations, are all brought together with a great deal of technical ingenuity in the form of control strategies and information flows.

This book focusses on the interface between modern renewable energy sources and classic conventional power generation. Any control of such a complex system must be based on knowledge of the current and anticipated input from the fluctuating sources; in this case, wind energy. Regulation of the conventional power station output and, in future, also increasing control of consumption must perform the feat of matching currently available power and demand. Today, forecasts of the anticipated output from wind energy, or energy weather forecasts, already play a significant role in controlling conventional power stations to minimise losses (and the associated CO_2 emissions) for providing regulating power. If a forecast can be made for a period of a few hours or days in advance, it is then possible to apply it together with a load forecast to adjust and optimise conventional power station scheduling to the expected wind power. This minimises fuel consumption, harmful emissions and costs.

The present book aims to review the current state of technology and the latest methods of wind power forecasting, sometimes also to lay out a vision for a coherent framework supporting the fundamental comprehension of the complexity involved. It takes us into a brand-new field of science called energy meteorology which draws on meteorology, physics, energy technology and energy economics. Especially, due to the challenges raised by the increasing share of renewable energy sources in electric power generation, this research field has grown significantly in importance and will become even more important in future. Wherever possible, the authors describe meteorological processes in a meteorological and physical way rather than limiting themselves to purely statistical input/output models. Apart from its practical advantages, this approach also has the huge benefit of quantitatively revising and improving our understanding of the basic processes involved, such as the development of atmospheric flows in the surface layer of the atmosphere.

Yet this work does not stop at theoretical considerations but also includes an extensive verification of the models developed so that the emerging processes can have direct practical effects. In that sense, the book represents one piece in the puzzle of one of the greatest challenges we face in the coming decades—the complete switchover from a conventional to a sustainable energy supply. Due to its high share in current electricity generation and the challenges it has created, wind energy helped to trigger and accelerate this development. It is clear that, approaching a sustainable electricity supply, wind energy is definitely not part of the problem, but part of the solution.

Oldenburg
July 2004

Hans-Peter Waldl
Jürgen Parisi

Preface

The combination of a scientifically challenging subject and the need to solve practical problems on the way to achieve a clean energy supply is our motivation to work in the field of wind energy. The market already demands for feasible solutions to efficiently integrate fluctuating wind power into the electrical supply system. Hence, in our view, wind power forecasting is a good example of how scientific methods can lead to innovations for the energy of tomorrow.

Wind energy is exciting for researchers as it is based on atmospheric physics, with a number of open questions. The non-linear nature of the atmosphere provides a rich variety of phenomena from small-scale turbulence to the formation of large-scale weather systems but typically prevents simple analytic solutions. Wind power predictions require knowledge in boundary-layer meteorology, flow simulation, time series analysis and modelling of the behaviour of wind farms to calculate the anticipated wind power production.

The physical approach to wind power predictions is the topic of our dissertations, in particular the description of thermal effects and of prediction uncertainty. As textbooks on wind energy typically contain little information concerning wind power prediction, we were asked to write this book based on our two theses. Though this origin is still noticeable, we intend to provide an overview of existing prediction methods in general and the physical approaches in particular. Since the field of wind power forecasting has developed impressively and dynamically in recent years and will continue to do so, this book is a first introduction and does not claim to give a complete review of all systems and methods that are available.

Our research is strongly embedded into the vivid wind power forecasting community in Europe. On various conferences and meetings over the years, we benefitted from the inspiring creativity and momentum of the people who work in this field. We are grateful that we had the opportunity to take part in European research projects on wind power prediction and we thank the European Commission for funding our work.

We would further like to thank the German Foundation for the Environment (DBU) for providing a grant to one of us (M.L.). Moreover, we thank Prof. Jürgen Parisi for his advice and support in writing the book, and Dr. Hans-Peter "Igor" Waldl who introduced us to the field of wind energy and supplied valuable comments on the manuscript. We thank the German Weather Service (DWD) for providing the prediction data and the Institut für Solare Energieversorgungstechnik (ISET) for making available the measurements data.

A big thank you goes to our wives for their solicitous support and their unlimited patience.

Oldenburg
June 2005

Matthias Lange
Ulrich Focken

Contents

1 Introduction 1
1.1 Purpose of This Book 2
1.2 Structure of the Book 2
1.3 Motivation for Wind Power Prediction 3

2 Overview of Wind Power Prediction Systems 7
2.1 Introduction 7
2.2 Numerical Weather Prediction 8
2.3 Statistical Systems 10
2.3.1 Wind Power Prediction Tool *WPPT* 11
2.3.2 Artificial Neural Networks 15
2.3.3 Fuzzy Logic 17
2.4 Physical Systems 17
2.4.1 Numerical Simulations 18
2.4.2 Diagnostic Models 21

3 Foundations of Physical Prediction Models 23
3.1 Introduction 23
3.2 Equations of Motion of the Atmosphere 25
3.2.1 The Navier–Stokes Equations 25
3.2.2 Reynolds-Averaged Equations 26
3.3 Physical Concepts of Boundary-Layer Flow 28
3.3.1 Eddy Viscosity and Mixing Length 28
3.3.2 Logarithmic Wind Speed Profile 31
3.4 Influence of Thermal Stratification on the Wind Profile 32
3.4.1 Description of Thermal Stratification 33
3.4.2 Corrections to the Logarithmic Profile 35
3.4.3 Geostrophic Wind as Driving Force 36

3.5 Conclusion ... 37

4 Physical Wind Power Prediction Systems ... 39
4.1 Introduction ... 39
4.2 Basic Scheme ... 39
4.3 Detailed Roughness ... 40
4.4 Thermal Stratification ... 43
4.4.1 Parametrisation of the Monin–Obukhov Length ... 44
4.5 Orography ... 47
4.5.1 Potential Flow ... 49
4.6 Farm Effects ... 50

5 Data ... 55
5.1 Introduction ... 55
5.2 Numerical Weather Predictions ... 56
5.3 Measurements ... 57
5.3.1 Wind Data and Power Output of Wind Farms ... 57
5.3.2 Meteorological Mast in Cabouw ... 57
5.3.3 Atmospheric Pressure ... 60

6 Assessment of the Prediction Accuracy ... 61
6.1 Introduction ... 61
6.2 Basic Visual Assessment ... 62
6.3 Distribution of Prediction Errors ... 64
6.3.1 Wind Speed Prediction ... 64
6.3.2 Error in Power Prediction ... 67
6.4 Statistical Error Measures ... 68
6.4.1 Decomposition of the Root Mean Square Error ... 69
6.4.2 Limits of Linear Correction Schemes ... 70
6.5 Wind Speed Prediction Error for Single Sites ... 74
6.6 Power Prediction Error for Single Sites ... 81
6.6.1 Comparison with Persistence ... 87
6.7 Conclusion ... 89

7 Correction of Wind Profiles Due to Thermal Stratification ... 91
7.1 Occurrence of Non-neutral Conditions ... 91
7.2 Application of Monin–Obukhov Theory ... 95
7.2.1 Correction of the Wind Profile up to 40 m ... 96
7.2.2 Correction of Wind Profile Above 40 m ... 98
7.2.3 Influence of Roughness on Stability Correction ... 101
7.2.4 Dependence of Stability Correction on Temperature Differences ... 101

7.3 Prediction of Thermal Stratification ... 105
7.4 Verification of the Stability Correction Scheme ... 107
7.4.1 Neuenkirchen ... 109
7.4.2 Hengsterholz ... 111
7.5 Conclusion ... 113

8 Assessment of Wind Speed Dependent Prediction Error ... 115
8.1 Idea Behind Detailed Error Assessment ... 115
8.2 Introduction of Conditional Probability Density Functions ... 116
8.2.1 Reconstructing the Distribution of the Power Prediction Error 117
8.3 Conditional Probability Density Functions of Wind Speed Data ... 120
8.4 Estimating the Distribution of the Power Prediction Error ... 126
8.5 Simple Modelling of the Power Prediction Error ... 129
8.6 Conclusion ... 132

9 Relating the Forecast Error to Meteorological Situations ... 135
9.1 Introduction ... 135
9.2 Methods from Synoptic Climatology ... 137
9.2.1 Principal Component Analysis ... 138
9.2.2 Cluster Analysis ... 140
9.2.3 Daily Forecast Error of Wind Speed ... 142
9.2.4 Tests of Statistical Significance ... 143
9.3 Results ... 143
9.3.1 Extraction of Climatological Modes ... 144
9.3.2 Meteorological Situations and Their Forecast Error ... 148
9.4 Conclusion ... 166

10 Smoothing Effects in Regional Power Prediction ... 169
10.1 Introduction ... 169
10.2 Ensembles of Measurement Sites ... 170
10.3 Model Ensembles ... 171
10.3.1 Distribution of German Wind Farms ... 176
10.4 Conclusion ... 177

11 Outlook ... 179

A Definition of Statistical Quantities ... 183
A.1 General Statistical Quantities ... 183
A.2 Error Measures ... 184

B Statistical Testing 187
B.1 The χ^2 Test 187
B.2 The Lilliefors Test 188
B.3 The *F* Test 189
B.4 *F* Test Results from Sect. 9.3 190

References 195

Index 203

1

Introduction

Wind energy is an important cornerstone of a non-polluting and sustainable electricity supply. Due to favourable regulatory frameworks this renewable energy source has experienced a tremendous growth in recent years, resulting in substantial shares of electricity produced by wind farms in the national energy mix of a number of countries. For example, during the last four years (1999–2003) the installed wind power capacity in Germany has risen from about 4.4 GW to more than 12.8 GW [3] and now covers approximately 4.7% of the total electricity consumption. And this is just the beginning, as from a global perspective wind energy has the potential to provide electric energy on an industrial scale. The European Wind Energy Association (EWEA) and Greenpeace estimate that by the year 2020 about 12% of the world's electricity production can be supplied with wind energy [75].

However, in practice, the integration of wind energy into the existing electricity supply system is a real challenge. A number of technical and economical objections against the large-scale utilisation of this renewable energy source have been brought forward, in particular by traditional energy suppliers and grid operators. In their view one major disadvantage of wind energy is that its availability mainly depends on meteorological conditions; hence, the power output of wind farms is determined by the prevailing wind speed and cannot be adjusted as conveniently as the electricity production of conventional power plants. As it is expensive to level out unforeseen fluctuations in the wind power production, grid operators and energy suppliers point out increasing costs due to wind energy.

Fortunately, wind power prediction systems which improve the technical and economical integration of wind energy into the electricity supply system are already available. They provide the information on how much wind power can be expected at a point in time in the next few days. Thus, they announce the variations in the electricity production of wind farms in advance and largely reduce the degree of randomness attributed to wind energy.

1.1 Purpose of This Book

This book is concerned with the short-term prediction of the power output of wind farms. It describes why wind power predictions are needed, how they are made and how forecast errors can be evaluated. A brief introduction to the broad field of wind power forecasting is provided giving an overview of the basic concepts behind different forecasting systems. The main focus of this book lies upon the properties of the so-called physical prediction models which are based on describing the physics of the atmospheric boundary layer. In particular, the influence of the thermal stratification of the atmosphere and the assessment of situation-dependent uncertainties of the prediction are discussed and quantitative models to describe these effects are presented.

This book is suitable as a textbook for scientists and students in the field of wind energy. It is also addressed to the entire wind energy community, in particular energy traders, energy dispatchers and consultants, as a beneficial reference book.

1.2 Structure of the Book

The book is organised as follows.

In Chap. 2 an overview of existing wind power prediction systems illustrates the different concepts of wind power forecasting. In particular, the two main approaches, namely statistical systems on the one hand and physical systems on the other, are described and their advantages and disadvantages are discussed. As most wind power prediction systems use large-scale numerical weather predictions (NWP) as input data, typical NWP models operated by different weather services are briefly described.

Chapter 3 provides the physical foundations of wind power predictions. Starting from the very fundamental equations of motion of the atmosphere, the basic concepts of boundary layer flow are derived in a concise manner. The logarithmic wind profile and its thermal corrections due to Monin–Obukov theory are discussed in detail.

The main part of this book is focused on physical prediction systems. In Chap. 4 the wind power prediction system *Previento*, which was developed at the University of Oldenburg, is used as an example to describe the transformation of the wind speed prediction of a numerical weather model into a forecast of the power output of wind farms. This includes the local refinement of the wind speed to the site-specific flow conditions considering a detailed roughness description and orographic effects as well as the modelling of the shadowing effects inside a wind farm.

The data basis used in the investigations of this book consists of measurements on the one hand and numerical predictions on the other. The origin and preprocessing of the two types of data are explained in Chap. 5.

The evaluation of the prediction error is an important aspect that is an essential part of this book. In Chap. 6 methods to assess the prediction accuracy are introduced and the relevance of understanding the statistical behaviour of the forecast error is shown. The forecast results for a number of test cases in northern Germany are discussed.

The thermal stratification of the lower atmosphere has a major influence on calculating the wind speed at hub height. Chapter 7 introduces a scheme to correct the logarithmic wind profile according to Monin–Obukhov theory. The benefits of the thermal correction are illustrated for two test cases.

In Chap. 8 the statistical properties of the forecast error are described in detail, introducing the idea of a situation-dependent assessment of the prediction uncertainty. In a first step the wind speed is considered, in particular the question whether the forecast accuracy depends on the magnitude of the wind speed. The important role of the non-linear power curve in amplifying initial errors in the wind speed forecast according to its local derivative is quantified, and a simple model to predict the specific forecast uncertainty is developed.

Moreover, in Chap. 9 a quantitative relation between the type of weather conditions and the corresponding accuracy of the wind speed prediction is derived. Methods from synoptic climatology are used to automatically classify weather situations based on a suitable set of meteorological variables. The typical prediction errors in each weather class are determined in order to find out whether dynamic low-pressure situations are really harder to predict than stable high-pressure types.

For practical purposes the combined power output of many spatially dispersed wind farms is of greater interest than that of a single one. Chapter 10 deals with regional smoothing effects and is an important step beyond the single-site perspective.

1.3 Motivation for Wind Power Prediction

In contrast to conventional power plants, the electricity production of wind farms almost entirely depends meteorological conditions, particularly the magnitude of the wind speed, which cannot directly be influenced by human intervention. Though this is a rather trivial fact, it makes a profound difference, technically as well as economically, in the way large amounts of wind energy can be integrated into electrical grids compared with the conventional sources. The so-called fluctuating or intermittent nature of the wind power production due to unforeseen variations of the wind conditions constitutes a new challenge for the players on the production and distribution side of the electricity supply system and is often used as an argument against the utilisation of wind power. Hence, the reputation and the value of wind energy would considerably increase if fluctuations in the production of wind power were known in advance. It is exactly the purpose of wind power prediction systems to provide this information. They are designed to produce a reliable forecast of the power output of wind farms in the near future so that wind energy can be efficiently integrated into the overall electricity supply.

Despite having rather different tasks and aims, the players on the liberalised energy market need to know about the anticipated consumption and production of electricity over a period of about 3 days in advance. Hence, energy traders, transmission system operators and power plant operators depend on forecasts of load as well as production, e.g., to make bids on the energy exchange or to schedule conventional power plants. Consequently, for them reliable wind power predictions are one important piece of information that leads to a cost-efficient energy supply with a large share of renewable energies.

In order to assess the benefits of wind power predictions in more detail and derive the boundary conditions for their operational use, some aspects of the electricity supply system will now be further explained from the point of view of a transmission system operator (TSO).

A secure electricity supply requires that at each point of time the electricity production match the demand as exactly as possible. It is the task of the TSO to carefully keep this balance. The load, i.e. the total consumption of electric power of households and industry, and its variations over the day are rather well known from experience and are expressed by the so-called load profiles. These daily load patterns are used to estimate the electricity demand of the next day with a relatively high accuracy.

In a world without wind energy the load profiles are sufficient for the TSO to work out a rather precise plan of how to satisfy the demand on a day-ahead basis. In a liberalised market environment the TSO basically has two options: produce electricity using its own power plants or buy electricity on the market. In the case of own production a schedule for the conventional power plants is made today that defines the number and type of power plants to be in operation tomorrow. Hence, the time horizon for the scheduling is about 48 h. This timetable considers the special characteristics of the different kinds of power plants, such as time constants to come into operation or fuel costs. If electricity is bought or sold on a day-to-day basis on the energy market, bids also have to be made about 48 h in advance. How the two options are combined depends on technical as well as economical considerations, e.g. those described by Poll et al. [94].

However, large shares of wind energy spoil this nicely established scheme to a certain degree, especially if wind power is unexpectedly fed into the system. From the point of view of the TSO, wind power acts as a negative load because the demand of electricity that has to be met by conventional power plants is reduced by the proportion of wind power available in the grid. This can be quite substantial in areas with high grid penetration of wind energy where the installed wind power is of the order of magnitude of the minimum load, which is, e.g., the case for certain areas in

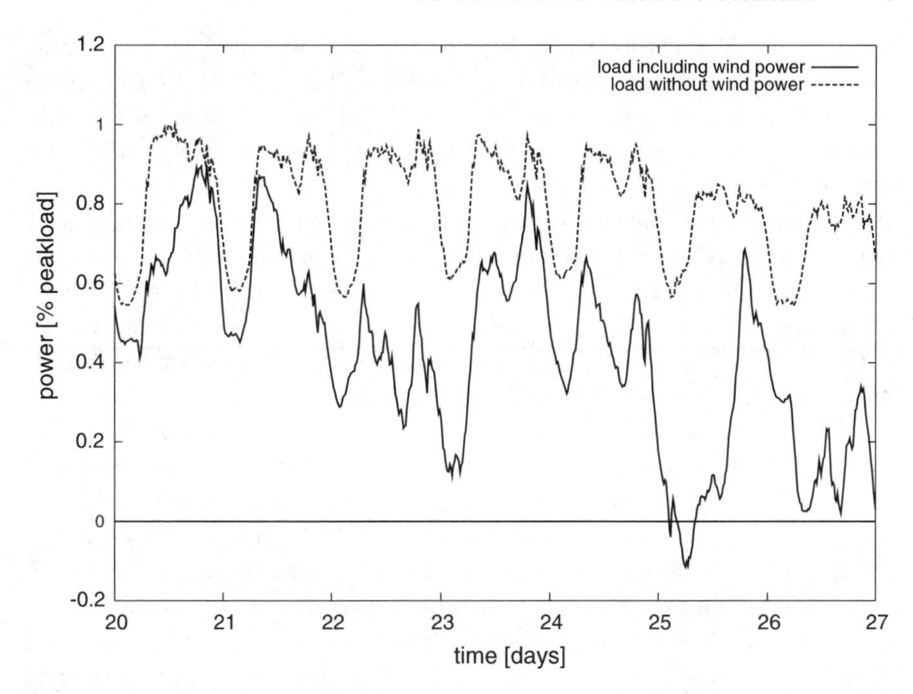

Fig. 1.1. Electrical load with and without wind power for a TSO with high grid penetration of wind energy for a period of 1 week. If no wind power is fed into the grid, the daily load pattern is very regular, reflecting the electricity demand of consumers and industry. In contrast to this, the contribution of wind power as a negative load leads to a rather fluctuating behaviour. In situations with high wind speeds the combined power output of wind farms can exceed the demand for a certain period of time (day 25)

northern Germany or Denmark. Figure 1.1 illustrates the effect of large amounts of wind power being fed into the electrical grid.

In Germany TSOs mainly deal with this situation by using additional balancing power (e.g. described by Tauber [109] and Dany [15]). Balancing power is generally applied to compensate for sudden deviations between load and production and can also be used to balance the fluctuating behaviour of wind power in the electrical grid. Keeping balancing power aims at being prepared for surprising situations, e.g. due to an unexpected drop in the power output of wind farms, which can be rather dramatic if many wind farms in a supply area switch themselves off for security reasons during a storm and the production decreases considerably. As surprises are not the kind of thing that are highly appreciated by TSOs, the amount of balancing power related to wind energy is relatively high and, therefore, expensive—a fact that is constantly pointed out by the TSOs; see e.g., [109]. Moreover, balancing power diminishes the environmental benefits of wind energy as it is technically realised by either making power plants operate with a reduced degree of efficiency or activating additional fossil fuel driven plants, as discussed by Leonhard et al. [64] and Dany [15]. Hence, the energy corresponding to balancing power that is additionally used to compensate for fluctuations in wind power has to be subtracted from the energy fed into the grid by wind farms, and this, unfortunately, reduces the amount of avoided CO_2 emissions.

As decisions regarding the scheduling of conventional power plants and placing bids on the electricity market are typically made in the morning for 1 day ahead, the wind power predictions have to cover a forecast horizon of at least 48 h. A sufficiently accurate prediction provides the decisive information concerning the availability of wind power over the next 1–2 days and, hence, allows for TSOs to efficiently integrate substantial shares of wind power into the existing electricity supply system.

The benefits are obvious because the amount of balancing power can be decreased if the prediction is reliable enough. Dany [15] found a quasi-linear relation between the forecast error and the need for balancing power caused by wind energy; thus, an improved prediction accuracy directly leads to a reduction in balancing power. The operational use of a reliable wind power prediction system not only enables TSOs to save money by using less balancing power, but the information on how much wind power will be available allows to trade wind power on the electricity market. It has already been illustrated for the Scandinavian situation by Mordhorst [78] that wind energy can be profitably traded on a liberalised market, in particular on the spot market, and Holttinen et al. [45] found that today's wind power prediction systems can already improve the income achieved by selling wind power under short-term market conditions. Hence, wind power forecasts increase the economic value of wind energy and help to make this renewable energy source competitive with conventional sources. In addition, the environmental advantage of wind energy is further increased as unnecessary CO_2 emissions due to balancing power are reduced.

2

Overview of Wind Power Prediction Systems

Abstract. This chapter provides an overview of existing wind power prediction systems and illustrates the different concepts of wind power forecasting. The two main approaches, statistical systems on the one hand and physical systems on the other, are described. Their advantages and disadvantages are briefly discussed. In addition, typical large-scale numerical weather prediction systems (NWP) which are mainly used as input for wind power predictions are introduced.

2.1 Introduction

In recent years a number of systems to predict the power output of wind farms have been developed. This chapter gives an overview of approaches to wind power prediction; the intention is to point out the different methods and concepts rather than providing a complete list of prediction models that are currently available. A recent overview can be found in [36].

The time horizon covered by the forecasts is given by the scheduling scheme of the conventional power plants and the bidding conditions on electricity markets which are typically of the order of one to two days ahead. Hence, prediction systems are required to provide the expected power output from 6 to at least 48 h, preferably in an hourly resolution.

The required time horizon is very important, as from a modelling point of view there is a fundamental difference between the so-called short-term predictions on time scales of a few days that are considered here and very short-term predictions in the range 0–3 h. While the longer time period is rather well described by numerical weather prediction systems which explicitly model the dynamics of the atmosphere, the very short range is typically dominated by persisting meteorological conditions where purely statistical approaches lead to better forecast results, e.g. [11, 79].

Most of the existing power prediction systems are based on the results of numerical weather prediction (NWP) systems. Hence, all the information about the future, in particular the expected evolution of the wind field, is provided by NWP systems. As NWP systems extrapolate the actual state of the atmosphere using the laws of physics, the accuracy of the numerical predictions over the desired time horizon is typically far better than any type of statistical or climatological approach which represent the average statistical behaviour.

The wind velocity vector, i.e. wind speed and wind direction, is, of course, the most important variable in terms of wind power prediction. It is the task of the wind power prediction system to convert this "raw information" typically given with a rather coarse spatial resolution by NWP systems into an adequate prediction of the power output of a wind farm. There are basically two approaches to transform the wind prediction into a power prediction. On the one hand, physical systems carry out the necessary refinement of the NWP wind to the on-site conditions by methods that are based on the physics of the lower atmospheric boundary layer. Using parametrisations of the wind profile or flow simulations, the wind speed at the hub height of the wind turbines is calculated. This wind speed is then plugged into the corresponding power curve to determine the power output. On the other hand, statistical systems in one or the other way approximate the relation between wind speed prediction and measured power output and generally do not use a pre-defined power curve. Hence, in contrast to physical systems the statistical systems need training input from measured data.

2.2 Numerical Weather Prediction

NWP systems simulate the development of the atmosphere by numerically integrating the non-linear equations of motions starting from the current atmospheric state. For this purpose the continuous real world has to be mapped on a discrete three-dimensional computational grid. This is not a spectacular step, but it is helpful to briefly reflect on the consequences.

Due to limited computer capacity the resolution of the numerical grid is finite, so that NWP systems cannot directly simulate processes on a sub-grid scale. This holds, in particular, for the influence of orographic structures of the terrain and localized thermal processes such as sea breezes. Where the sub-grid processes have an influence on the evolution on a larger scale, e.g. turbulence or the formation of clouds, these phenomena have to be parametrised, which means that their macroscopic effect is modeled without considering the microscopic details. Thus, the NWP models account for the atmospheric behaviour from a large scale, i.e. synoptic weather systems with about 1000-km extension, down to a scale of the order of 10 km.

In addition, the finite resolution means that the variables calculated at each grid point represent averages over the corresponding grid cell. Hence, the predicted value might not be optimal for all locations inside one grid cell. Hence, care has to be

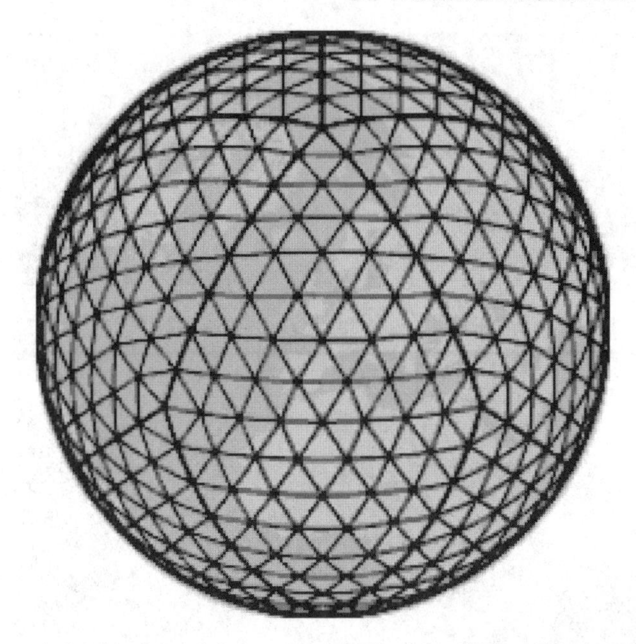

Fig. 2.1. Numerical grid of the global model (GME) of the German Weather Service with a spatial resolution of 60×60 km^2

taken if predictions and point measurements are compared. The measurement values normally have to be time-averaged to be comparable with the forecast.

For both types of wind power prediction systems, NWP provides the necessary input. Due to their complexity and the large amount of data collection that is required to run these models, they are typically operated by national weather services. In general, the weather services use one global model with a horizontal resolution ranging from 100×100 km^2 down to about 50×50 km^2 to capture the worldwide development of the weather systems. Figure 2.1 shows for example the grid of the global model (GME) of the German Weather Service (DWD). Such a global model then drives a local model with a higher spatial resolution from 50×50 km^2 down to 7×7 km^2 (e.g. the *Lokalmodell*) but with a smaller domain which is typically centred around the home country of the weather service (see Fig. 2.2).

Setting the initial state for a prediction run requires a large amount of data collection. The atmospheric state is measured on a regular basis by a large number of synoptic stations, buoys, radio sondes, ships, satellites and planes all over the world. As data formats and measuring cycles are standardized by the World Meteorological Organisation (WMO), the global meteorological data is available to the weather services to set the initial conditions for their NWP models.

Table 2.1 gives an overview of operational NWP systems operated by the European weather services.

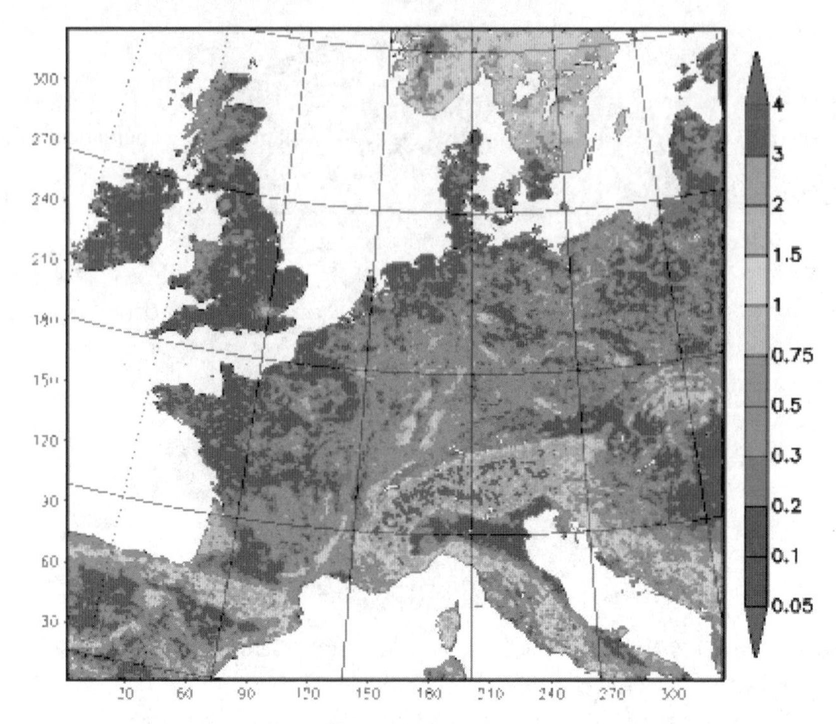

Fig. 2.2. Domain of the *Lokalmodell* (LM) of the German Weather Service with a spatial resolution of $7 \times 7\ \text{km}^2$

Table 2.1. Overview of operational NWP systems in Europe

Model name	Weather service	Horizontal resolution (km)
Globalmodell (GME)	German Weather Service (DWD)	60
Lokalmodell (LM) [17, 18]		7
Arpege	Meteo France	19–250
Aladin [73]		9.5
Hirlam [56, 99]	Sweden, Denmark, Norway, Spain, Iceland, Ireland, Finland, the Netherlands	10–50
ECMWF [32]	European Center for Medium Range Forecasts	40
Unified Model [74]	MetOffice, UK	60/11

2.3 Statistical Systems

The focus of this book lies on physical systems for wind power prediction. However, in the following, a selection of statistical systems and their properties are briefly described to point out how different this approach is.

2.3.1 Wind Power Prediction Tool *WPPT*

A statistical system that has been used operationally in Denmark since 1997 was developed in co-operation between the Institute of Informatics and Mathematical Modelling (IMM) at the Technical University of Denmark, the Danish power production utility Elsam and Eltra, the transmission system operator TSO in the western part of Denmark. The system, called WPPT, derives the relation between predicted wind speed and measured power output without an explicit local refinement. The optimal weights between the online measurements and the meteorological forecast are continuously re-calculated. This has the advantage that the parameters are automatically adapted to long-term changes in the conditions, e.g. variations in roughness due to seasonal effects or model changes in the NWP system.

WPPT consists of several advanced mathematical models which are combined to generate the prediction of the aggregated power output of wind farms in a certain area. It is beyond the scope of this book to cover all these models in detail. Further descriptions of the system can be found in [67] and [82, 83, 85].

Recursive Least Squares Method

Mathematically, WPPT is based on an ARX (Auto-Regressive with eXogeneous input) type model. Under this approach the power production is described as a non-linear and time-varying (and, hence, non-stationary) stochastic process which, therefore, reflects basic properties of the underlying dynamics of the atmosphere. Models describing this kind of stochastic processes cannot be handled successfully using ordinary least squares estimation techniques. In order to cover the non-linear and non-stationary behaviour, adaptive estimation techniques are required which are based on recursive least squares (RLS) estimation. This method makes the system self-calibrating and allows WPPT to compensate for slow changes.

In the following, the statistical method behind WPPT (Fig. 2.3) is briefly described outlining first the general mathematical approach and then the specific model that is used to predict wind power.

In general, the regression method requires a stochastic model that is linear in the parameters such that the expected value of the current observation y_t at time t is given by

$$E(y_t) = \vec{x}_t^T \cdot \vec{\Theta} \tag{2.1}$$

where $E(.)$ denotes the expectation value, $\vec{\Theta}$ is the parameter vector to be determined and $\vec{x}_t$ is the vector of independent variables fed into the system.

To actually obtain the parameters in terms of an RLS method, a criterion that allows to determine the parameters is necessary. In ordinary linear regression this typically involves all data points that are available, i.e. the measurements and predictions that have been recorded so far enter the estimation procedure with the same

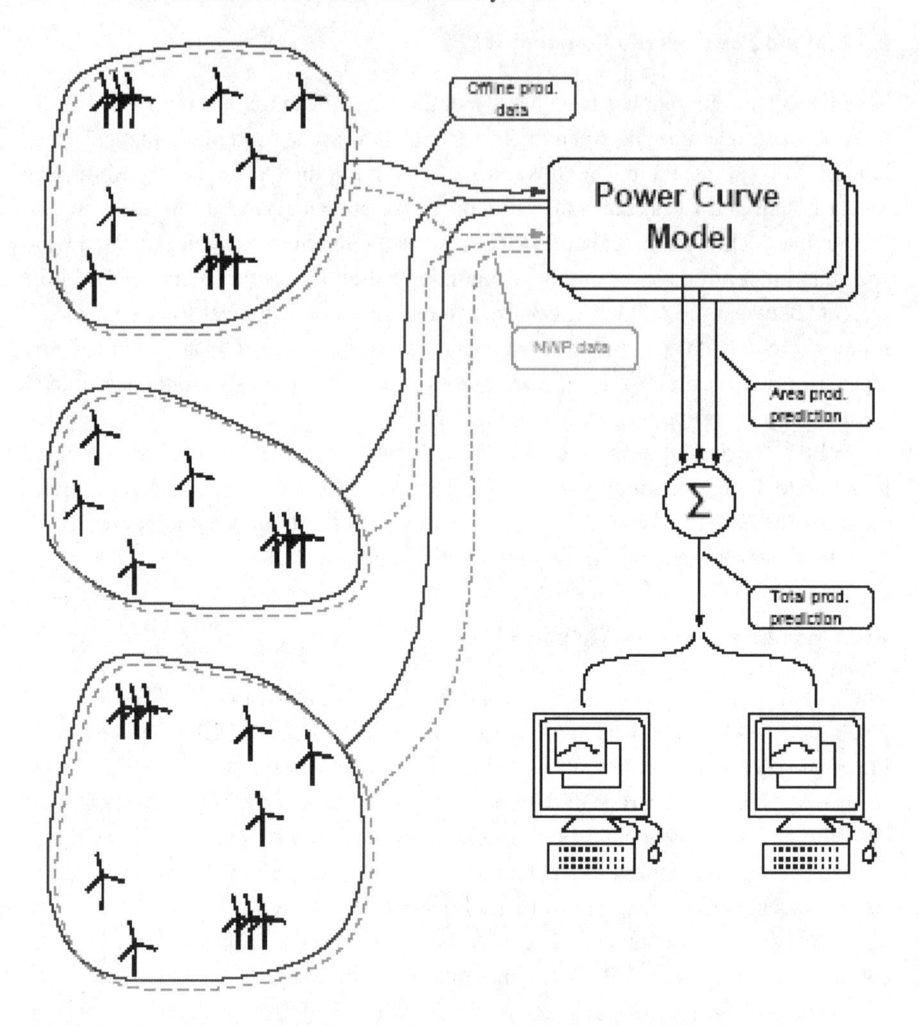

Fig. 2.3. General scheme of the wind power prediction tool WPPT by IMM. The system considers the characteristics of a large number of wind farms or stand-alone wind turbines. The power curves are derived from measured data of the power production. These data are either obtained online for several representative sites or offline from a large number of wind turbines. Picture provided by IMM

weight and in this case the parameters $\vec{\Theta}$ are chosen such that the mean squared difference between observations $y_s (s = 1, \ldots, t)$ and prediction $\vec{x}_s^T \cdot \vec{\Theta}$ is minimised; hence,

$$\sum_{s=1}^{t} \left(\vec{x}_s^T \cdot \vec{\Theta} - y_s \right)^2 \rightarrow \min \ . \tag{2.2}$$

As shown in [65], the solution of (2.2) is given by

$$\hat{\vec{\Theta}} = (X^T X)^{-1} X^T \vec{y} \,, \tag{2.3}$$

where the matrix X^T is defined by

$$X^T = [\vec{x}_1 \cdots \vec{x}_t] \tag{2.4}$$

and the vector $\vec{y}$ is given by

$$\vec{y} = [y_1 \cdots y_t]^T \,. \tag{2.5}$$

The algorithm is formulated recursively to avoid, in online operation, the update of the parameters involving all data points of the past. The aim is to find a new estimate $\hat{\vec{\Theta}}_t$ at time t based on the new values and the previous estimate $\hat{\vec{\Theta}}_{t-1}$.

With the abbreviations

$$R_t = \sum_{s=1}^{t} \vec{x}_s \vec{x}_s^{\,T} \qquad \text{and} \qquad \vec{f}_t = \sum_{s=1}^{t} \vec{x}_s y_s \tag{2.6}$$

(note that $\vec{x}_s \vec{x}_s^{\,T}$ is a dyadic product) the RLS algorithm can be summarised as

$$\begin{aligned} R_t &= R_{t-1} + \vec{x}_t \vec{x}_t^{\,T} \,, \\ \hat{\vec{\Theta}}_t &= \hat{\vec{\Theta}}_{t-1} + R_t^{-1} \vec{x}_t \,, \left[y_t - \vec{x}_t \cdot \hat{\vec{\Theta}}_{t-1} \right] \end{aligned} \tag{2.7}$$

and describes the update of the parameters $\hat{\vec{\Theta}}$ from time $t-1$ to t. Traditionally, the initial parameter setting is $\vec{\Theta} = 0$ and $R = 0$.

However, in order to make the system account for variations in the parameters, the influence of data points from the past has to be dampened using a forgetting factor. Hence, the minimisation criterion for RLS is formulated as

$$\sum_{s=1}^{t} \lambda^{t-s} \left(\vec{x}_s^{\,T} \cdot \vec{\Theta} - y_s \right)^2 \to \min \,, \tag{2.8}$$

where λ is the forgetting factor, with $0 < \lambda \leq 1$. Hence, the adaptivity of the parameters is obtained by multiplying the older values with an exponentially decreasing weight function.

The choice of the forgetting factor λ is a compromise between the ability to track time-varying parameters and the noise sensitivity of the estimate, i.e. the memory of the algorithm must be long enough to provide a number of data points for a reliable statistic but still sufficiently short to follow variations. Typically, λ is in the range $0.95 < \lambda < 0.999$. If λ is close to 1 the number of effective data points that are used is given as

$$N_{\text{eff}} = \frac{1}{1-\lambda} \,. \tag{2.9}$$

The forgetting factor can easiliy be included into the RLS procedure in (2.7) and, finally, the adaptive least squares algorithm is given by

$$R_t = \lambda R_{t-1} + \vec{x}_t \vec{x}_t^{\,T} ,$$
$$\hat{\vec{\Theta}}_t = \hat{\vec{\Theta}}_{t-1} + R_t^{-1} \vec{x}_t \left[y_t - \vec{x}_t \cdot \hat{\vec{\Theta}}_{t-1} \right] . \tag{2.10}$$

If the prediction step is 1, the estimate of $\hat{\Theta}$ at time $t-1$ is used to construct the prediction value $\hat{y}_{t|t-1}$ that is valid for time t based on the state of the system of time $t-1$, i.e.

$$\hat{y}_{t|t-1} = \vec{x}_t^{\,T} \cdot \hat{\vec{\Theta}}_{t-1} , \tag{2.11}$$

involving the 1-step prediction errors $\left[y_t - \vec{x}_t \cdot \hat{\vec{\Theta}}_{t-1} \right]$.

For a k-step prediction, pseudo prediction errors are used:

$$\tilde{y}_{t|t-k}^{\text{pseudo}} = y_t - \vec{x}_{t-k}^{\,T} \hat{\vec{\Theta}}_{t-1} , \tag{2.12}$$

which involves the variables known at time $t-k$ (i.e. $\vec{x}_{t-k}$) and the most recent parameter estimate $\hat{\vec{\Theta}}_{t-1}$.

The true k-step prediction from time t to $t+k$ is then calculated using the pseudo-error given by (2.12) together with the iteration (2.10), and one obtains

$$\hat{y}_{t+k|t} = \vec{x}_t^{\,T} \hat{\vec{\Theta}}_t . \tag{2.13}$$

Power Prediction

The model described above is now used to predict the power output of a wind farm based on measurements of wind speed as well as power output and meteorological forecasts of wind speed.

In general, a wide variety of different models is possible due to the many combinations of variables. One rather successful model is called WPPT2 and is formulated as

$$\begin{aligned}\sqrt{p_{t+k}} = \; & a_1 \sqrt{p_t} + a_2 \sqrt{p_{t-1}} , \\ & + b_1^0 \sqrt{w_t} + b_2^0 w_t + b_1^m \sqrt{w_{t+k|t}^m} + b_2^m w_{t+k|t}^m , \\ & \sum_{j=1}^{3} \left[c_j^c \cos\left(\frac{j 2\pi h_{t+k}^2 4}{24} \right) + c_j^s \sin\left(\frac{j 2\pi h_{t+k}^2 4}{24} \right) \right] + m + e_{t+k} , \end{aligned} \tag{2.14}$$

where p_{t+k} is the desired predicted power at time $t+k$, p_t is the observed power at time t, w_t is the observed wind speed at time t, $w_{t+k|t}^m$ is the predicted wind speed at $t+k$ available at time t, h_{t+k}^{24} is the time of day at time $t+k$, e_{t+k} is a noise term, and

$a_1, a_2, b_1^0, b_2^0, b_1^m, b_2^m, c_j^c, c_j^s$ ($j = 1, \ldots, 3$) and m are the time-varying parameters to be estimated. An applicable choice of the forgetting factor is $\lambda = 0.999$.

The square root terms, e.g. $\sqrt{p_{t+k}}$, are introduced to account for the non-Gaussian distribution of the power, while the sine–cosine terms model the diurnal variation over the day.

Nielsen et al. [84] thoroughly investigated different models of the ARX type. They found that meteorological wind speeds are necessary for an adequate performance for all prediction times. Moreover, for shorter prediction horizons the models rely on observed values of power and to a smaller extent on wind speed, but as the prediction time increases the terms describing diurnal variation become more important. For forecast horizons larger than 18 h, the impact of observations is negligible.

WPPT is an example of a statistical power prediction system and, hence, from a different class than the systems that are based on a physical approach, which are the main topic of this book. The statistical systems are based on a "learning" algorithm, which means that they are designed to implicitly model the relation between meteorological variables and measured power output. Hence, physical parameterisations are in general not included (an exception would be the diurnal variation in (2.14)).

Clearly, the advantage of statistical systems is that they are mostly self-calibrating to inherent changes in the system and that they incorporate site-specific conditions automatically. On the other hand, one disadvantage lies in the requirement for online measurement data for a large number of wind farms. In addition, expertise concerning the physics of the atmosphere is still necessary to select the meteorological variables that are used in the learning process, e.g. thermal stratification of the atmosphere.

For use in complex terrain in Spain, WPPT has been modified by Marti et al. [70] using the Spanish version of HIRLAM as input. However, as both the type of terrain and the climate of the Iberian peninsula are rather challenging, a new model chain, called *LocalPred* [68, 69], is being developed that intends to combine statistical time series forecasting and high-resolution physical modelling based on meso-scale models.

2.3.2 Artificial Neural Networks

A system for online monitoring and prediction of wind power that is in operational use at several German TSOs has been developed by the Institut für Solare Energieversorgungstechnik (ISET) in Kassel and is called WPMS [21, 98]. The system provides an online estimation of the wind power that is currently fed into the electrical grid based on extrapolating measurements at representative wind farms. In addition, WPMS contains a wind power prediction tool that uses artificial neural networks (ANNs) to represent the relation between meteorological prediction and power output.

The prediction system is designed to provide forecasts of the aggregated power output of many wind farms in a region with a time horizon of 0–72 h. First, for a number of representative wind farms the numerical predictions of wind speed and further meteorological variables of the German weather service are transformed into a wind power forecast by an ANN which has been trained with historical data of that site. Then the predicted power outputs of the representative sites is scaled up and accumulated to obtain the aggregated power of the desired region. In addition, online measurement data of the representative sites are used to improve the forecast for the short forecast range of 1–8 h.

The concept of ANNs is originally derived from biological nerve cells, the so-called neurons, which are interconnected to a high degree and able to generate a specific response depending on the activity level of neighbouring cells. ANNs are a wide class of mathematical models which partly emulate the behaviour of these biological systems. The activity level of an artificial neuron is determined by integrating incoming signals from other neurons. Hence, in mathematical terms its activity is described by a weighted sum. Let $s_i(t)$ be the activity of the ith neuron at time t; then the activity of a neuron j which is connected to N neurons, s_i, is described by

$$s_j(t) = \sum_{i=1}^{N} w_{ij} s_i(t) \tag{2.15}$$

where w_{ij} are pairwise weighting factors between the neurons. Note that the w_{ij} are real numbers which can, in particular, be negative if the neuron i is inhibiting the activity of neuron j. The weighting factors are important parameters of an ANN. They are modified according to certain rules during the training period and determine the behaviour of the ANN.

Figure 2.4 shows the basic scheme of the ANN that is used in WPMS. It consists of different layers which are responsible for the perception of input, the processing according to the previous training and the output of the power prediction. Hence, the basic idea behind the system is to relate the meteorological forecasts to the corresponding measured power output in order to create a multi-dimensional power curve for each representative site. The advantage of this over using certified power curves is clearly that the individually trained power curves partly compensate for systematic errors. One strength of ANN is that it is able to deal with new input, e.g. forecast situations that have not been trained, or erroneous input, e.g. due to incomplete data.

Details regarding neural networks in general can be found, e.g., in [125]. The layout of the ANN used in the framework of this prediction tool and the methods to train the weighting factors are fully described by Ernst [20] and Rohrig [97].

2.3.3 Fuzzy Logic

A different statistical approach involving prediction models based on fuzzy logics has been developed by Ecole de Mines (ARMINES), France, to provide a short-term prediction of wind power. The training is based on NWP data on the one hand and online SCADA measurements on the other. An earlier version of the system is called *MORE-CARE* and is operationally used in Ireland, Crete and Madeira (Pinson et al. [93] and Kariniotakis et al. [54]).

2.4 Physical Systems

Physical systems use the concepts of atmospheric dynamics and boundary-layer meteorology to carry out the spatial refinement of the coarse output of NWP systems to the specific on-site conditions as well as the transformation of the predicted wind speed to the hub height of the wind turbines. Two basic classes of physical prediction systems can be distinguished: first, models which are based on operational fluid dynamical simulations similar to those of NWP systems, and, second, the diagnostic models which mainly use parametrisations of the boundary layer.

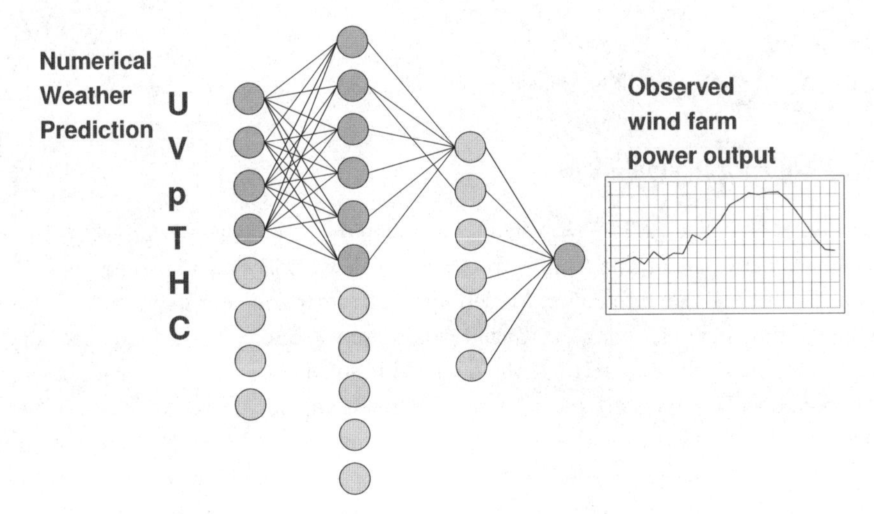

Fig. 2.4. Basic scheme of the ANN used in WPMS. The different layers are responsible for the perception of input, the processing according to the previous training and the output of the power prediction. The lines indicate the connections between the neurons (large dots) and the weighting factors. Picture provided by ISET

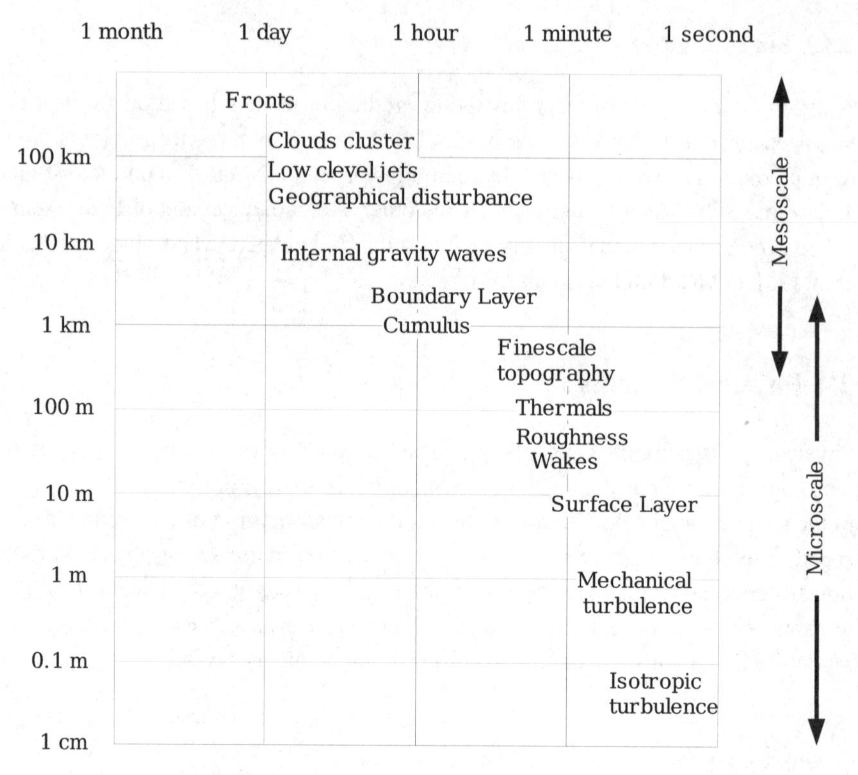

Fig. 2.5. Typical orders of magnitude in time and space for atmospheric phenomena [86]

2.4.1 Numerical Simulations

As discussed in Sect. 2.2 NWP systems do not explicitly simulate the complete range of atmospheric phenomena ranging from large-scale weather systems to energy dissipation in micro-scale turbulence. Modelling all effects would consume too much calculation time in operational service and would produce more details than necessary for a specific application. Hence, the numerical models that describe the dynamics of the atmosphere typically focus on certain effects and can, therefore, be classified according to their spatial and temporal scale. This is illustrated in Fig. 2.5, where the length and time scales of typical atmospheric phenomena are shown. It is important to note that the numerical models do not differ only in their spatial resolution but also in the set of equations of motions which is also adapted to the kind of physical effects that have to be described on the specific scale.

Basically, the numerical models are divided into three classes:

- NWP models: 1000 km down to 10 km
- Meso-scale models: 10 km down to 1 km
- Micro-scale models: several 100 m down to 0.01 m

NWP models are described in more detail in Sect. 2.2.

Meso-scale Models

The meso-scale models are driven by large-scale NWP models, i.e. the NWP models provide the boundary conditions. This process is called nesting, where the idea is to use a high-resolution model only in areas of special interest to cover atmospheric effects on a smaller scale that cannot be resolved by the coarser model. Mostly, meso-scale models are used for three to four nesting operations, because the nested grid domain should be at least one third the size of the bigger domain.

In contrast to large-scale NWP models the meso-scale models are typically non-hydrostatic, which means that they can directly model vertical motions, e.g. due to buoyancy. In addition, the parameterisation of the planetary boundary layer is adopted to the higher spatial resolution and the turbulence-closure schemes are modified. In particular, the advantage of meso-scale models lies in a more detailed consideration of thermal effects, which is very beneficial in terrain with high solar irradiation and near the coast to capture sea breezes. Meso-scale models are also important for complex terrain, e.g. in Spain or Norway, where wind speed and wind direction are locally inhomogeneous. An overview of meso-scale modelling can be found in [89]. For wind energy applications the well-documented meso-scale model MM5 [38] is widely used, so are the models GESIMA [53,71], KLIMM [22], RAMS [90, 123], Fitnah [39,40], and the Eta Model [72].

Micro-scale Models

In order to account for flow effects in the direct vicinity of a wind farm, micro-scale models with different degrees of complexity are used.

The rather popular class of mass-consistent models (e.g. AIOLOS [29], WIEN [104], NOABL [111], MesoMap [4], *Winds* [33]) is only based on the principle of conservation of mass, which is equivalent to a divergence-free wind field. Hence, frictional and thermal effects are typically ignored or added afterwards as a correction. A well-known model of this type is the Wind Atlas Analysis and Application Program (WAsP) model that is described in more detail in Sect. 4.5. Mass-consistent models are able to model the effect of slightly complex terrain but tend to overestimate horizontal wind speed gradients if the slope of the terrain is too large. The flow solutions typically have certain symmetries such that they can be computed very time efficiently. As the initial flow field has a large impact on the results, the influence of direction-dependent roughness can easily be implemented (c.f. [29]).

More advanced models are based on the Navier–Stokes equations, e.g. Phoenics [41]. They consider the full flow problem but disregard thermal effects. Most of these models are not developed for the atmospheric boundary layer, but, e.g., for pipe flow or the aerodynamics of rotor blades. Therefore, specific parameterisations for the boundary layer must be added. An overview of these models can be found in [25]. These models can be applied in a stationary mode or as prognostic models like meso-scale models. But they are currently not operationally used for prognostic wind power prediction due to limited computer resources.

Applications

In terms of wind power predictions, numerical models are used in two different modes: first, the dynamical mode where each operational prediction is calculated by solving the equations of motion for the actual flow situation, and second, the offline mode where the numerical models are used to generate a look-up table for a number of standard flow situations which e.g. can be determined by a cluster analysis (see Chap. 9); in this mode the dynamical calculations are carried out only once and the results are later used to correct the flow.

A prediction system in dynamical mode is *eWind*, which has been introduced by Bailey et al. [4]. It is a prediction system that is physical in terms of local refinement of the wind conditions but statistical in terms of determining the power output. It uses the meso-scale atmospheric model MASS 6, which is similar to MM5. It is driven by a regional NWP with a coarser resolution. Hence, eWind actually simulates the local flow instead of using a parametrisation of the wind profile. However, adaptive statistical methods which require on-site measurement data are used in the last step to translate wind speed into power and to correct for systematic errors in the prediction. Another operational forecasting system based on MM5 is implemented at CENER in Spain. It runs with a four-nested domain driven by the American AVN model [69].

Instead of using the NWP wind speed forecast as given input and modify it afterwards for wind power applications, Jørgensen et al. [50] follow a more general approach by implementing wind power forecasts directly into the large-scale weather prediction system HIRLAM. Again, nesting is used to increase the spatial resolution of the NWP in order to optimise the system with regard to accurate wind speed forecasts. As it is directly related to the NWP the power prediction system can take full advantage of the complete set of meteorological variables provided by the weather model in its internal temporal resolution which is typically much higher than the usual time steps of 1 h provided to customers.

One example for the look-up table approach is considering the flow over complex terrain in the WAsP model (Sect. 4.5), where the effect of a hill on the unperturbed flow is calculated with a mass-consistent model for several directions and wind speeds. The speed-up effects are expressed as factors in a look-up table and used to correct the actual flow situation. The look-up table approach is also applied in combination with statistical models, e.g. an early version of the system WPMS by ISET (Sect. 2.3.2) used the meso-scale model KLIMM [22] in that way.

2.4.2 Diagnostic Models

Diagnostic models are based on parameterisations of the boundary layer flow without further dynamical situations. The parameterisations used in the context of wind power prediction are derived from basic physical principles in combination with experimental observations (details can be found in Chap. 3).

One of the first physical power prediction systems with a prediction horizon up to 72 h was developed by Landberg [58, 59] at the National Laboratory in Risø in 1993. The procedure is based on a local refinement of the wind speed prediction of the NWP system HIRLAM [99] operated by the Danish Meteorological Institute. The refinement method to adapt the NWP output to the local conditions at the site was derived from techniques that had been used before for wind potential assessment in the framework of the European Wind Atlas [115]. Local surface roughness, orography describing the hilliness of the terrain, obstacles, and thermal stratification of the atmosphere are taken into consideration, leading to forecast results that were significantly better than persistence, i.e. the assumption that the current value will also be valid in the future. The prediction system has been commercialised under the name *Prediktor* for operational use.

The power prediction system *Previento* has been developed at the University of Oldenburg [7,77] and is in operational use in Germany. It is based on the same principle as Prediktor in terms of refining the prediction of wind speed and wind direction taken from the NWP system. The local conditions are derived by considering the effects of the direction-dependent surface roughness, orographic effects and, in particular, atmospheric stability [27, 28] on the wind profile. Moreover, the shadowing effects occurring in wind farms are taken into account. If measurement data from the wind farms are available, a systematic statistical correction of forecast errors is applied. As in practice the combined power output of many spatially dispersed wind farms in a region is of greater interest than that of a single wind farm, *Previento* contains an advanced up-scaling algorithm that determines the expected power output of all wind farms in a certain area based on a number of representative sites selected in an appropriate manner [30]. In addition to the power prediction itself, *Previento* also provides an estimate of the uncertainty of the specific forecast value [31] to allow users an assessment of the risk of relying on the prediction. For regional up-scaling and uncertainty estimates, a part of the results of this work has been implemented into *Previento* and will be described in more detail in Chaps. 9 and 10.

In general, the development of physical systems requires expert knowledge of meteorology and boundary layer physics. The complex phenomena of the atmosphere can either be dynamically simulated or described by parametrisations. Hence, the physical approach to wind power prediction is like open-heart surgery, where a number of detailed models have to be developed and put together in the end to obtain a useful model to transfer meteorological predictions into expected power output. Thus, in contrast to statistical systems, which are black boxes with regard to meteorology, the physical systems clearly have the advantage that individual atmospheric processes can be systematically investigated to optimise the models to describe them.

However, physical prediction systems are sensitive to systematic errors due to erroneous initial information. For example, if the local surface roughness is too large, the refined wind forecast underestimates the real situation; or if the power curve of a specific wind turbine does not correspond to the certified power curve implemented in the prediction model, the power prediction is systematically wrong. Consequently, though physical systems can in principle work without measurement data, the prediction accuracy is generally improved if measured wind speed and power output are available to calibrate the forecast.

3

Foundations of Physical Prediction Models

Abstract. This chapter provides the physical foundations of describing atmospheric wind fields in the framework of physical wind power prediction models. The equations of motion are introduced and the dominating physical effects in different regimes of the atmosphere are discussed, leading to the definitions of surface layer, Ekman layer and geostrophic wind. The physical parametrisation of the vertical wind profile is derived from basic concepts of boundary-layer meteorology. To include the effects of thermal stratification on the wind profile the basic concepts of Monin–Obukhov theory are shown.

3.1 Introduction

The physics of the atmosphere is rather complex as it is a nonlinear system with an infinite number of degrees of freedom. The dynamics can be described in mathematical terms by equations of motion that are derived from the principles of conservation of mass, momentum and heat. But the non-linear structure of these fundamental equations does in general not allow for analytic solutions such that approximate solutions for realistic states of the atmosphere can in most cases only be obtained under simplifying assumptions.

In the context of wind energy applications, two different approaches of describing the atmosphere are important. First, a forecast of the dynamical behaviour of the atmosphere and, in particular, the wind fields has to be obtained by numerically solving the equations of motion. These numerical weather predictions (NWP) provide the input for the power prediction system in a rather coarse spatial resolution. Hence, secondly, simpler models to describe the local wind conditions at a site of interest are needed to adapt the NWP forecast to the local flow conditions.

Before the details of the two approaches are explained in the following sections, the general ideas of dealing with the complex behaviour of the atmospheric flow are briefly shown. To circumvent the difficulties of solving the full set of equations, the atmosphere is divided into several horizontal layers to separate different flow

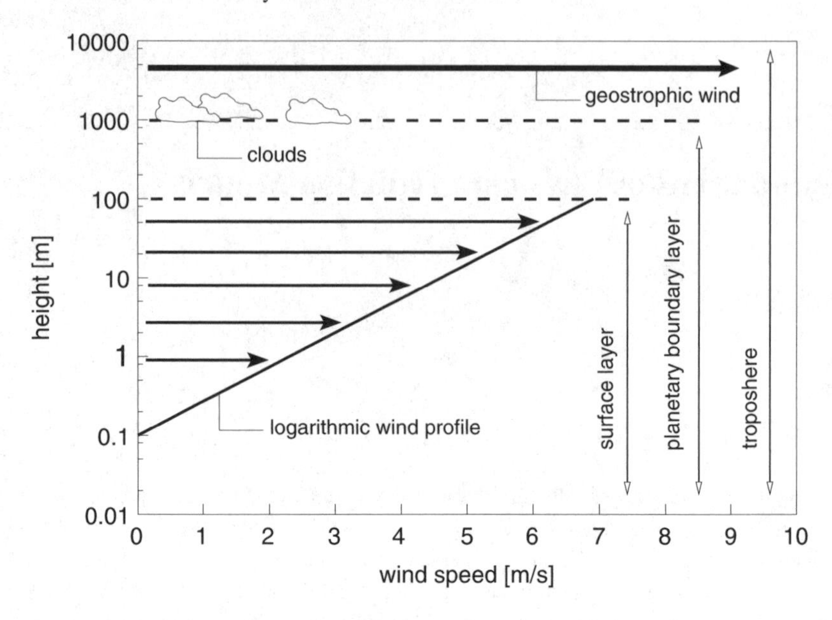

Fig. 3.1. Schematic illustration of the horizontal layers in the troposphere which comprises the lower part of the atmosphere that is important for wind energy. In the top layer above approximately 1 km, there is little influence of the ground and the wind is geostrophic, driving the flow in the lower layers. The wind field in the surface layer (up to about 100 m) is mainly dominated by friction exerted by the ground and turbulent mixing. Under certain assumptions a logarithmic profile can be derived for the wind speed

regimes. These layers are defined by the dominating physical effects that influence the dynamics. For wind energy use, the troposphere which spans the first five to ten kilometres above the ground has to be considered as it contains the relevant wind field regimes as illustrated in Fig. 3.1.

Heights greater than approximately 1 km are the domain of large-scale synoptic pressure systems, i.e. the well-known highs and lows. Their wind field is largely dominated by the horizontal gradients of pressure and temperature as well as the Coriolis force, caused by the rotation of the earth. As the influence of the ground is rather weak, Coriolis force and pressure gradient force can typically be considered as balanced, leading to the geostrophic wind that blows parallel to the isobars.

The geostrophic wind field is regarded as the main driving force of the flow in the underlying atmospheric layer denoted as planetary boundary layer (PBL). The wind in the PBL is dominated by the influence of the friction exerted by the earth's surface. Typically, the flow near the surface is turbulent, which provides a highly effective coupling mechanism between wind speeds at different heights, leading, in particular, to vertical transport of horizontal momentum that is directed towards the ground where the wind speed has to vanish. This momentum flux based on turbulent mixing is by far larger than it would be when based on molecular viscosity alone, so that the change of the surface wind with height strongly depends on the degree of turbulence in the atmosphere.

Considering in mind the qualitative picture of large-scale geostrophic wind fields that drive the surface winds by an effective turbulent coupling mechanism, the following sections provide a more quantitative understanding of the general properties of atmospheric dynamics. The fundamental mathematical description needed for wind power predictions is concisely presented, concentrating on the air flow in the different layers of the lower atmosphere. At first only pure flow effects are described leading to an approximate analytic expression for the vertical wind profile. Then the important thermal effects are included as corrections to this profile. For further reading a comprehensive and detailed discussion of the theoretical concepts of the atmosphere can be found, e.g., in Arya [2] and Stull [107]. Basic fluid dynamics is given, e.g., in Tritton [112].

3.2 Equations of Motion of the Atmosphere

3.2.1 The Navier–Stokes Equations

A fluid element in the atmosphere is exposed to external forces on the one hand and internal forces on the other hand. The external forces are the pressure gradient force, the Coriolis force and gravity while internal forces are caused by molecular friction between fluid elements. The difference between high and low pressure with the corresponding force directed from high to low constitutes the pressure gradient. Gravity is, of course, directed towards the centre of the earth and independent of the state of the flow while the Coriolis force which is due to the rotation of the earth is proportional to the speed of the fluid element and acts perpendicular to the direction of motion. As air is a Newtonian fluid the molecular friction is assumed to be proportional to the local velocity gradient, so that the molecular viscosity, η, is solely a property of the fluid and does not depend on the state of the flow.

With the basic principle of the conservation of momentum the dynamics of the flow field can be derived by equating the rate of change of momentum to the forces that are exerted on a fluid particle. This leads to the Navier–Stokes equation of the atmosphere given as

$$\underbrace{\rho\left(\partial_t \vec{u} + (\vec{u}\cdot\nabla)\,\vec{u}\right)}_{\text{change of momentum}} = \underbrace{-\nabla p}_{\text{pressure gradient}} \underbrace{+\,2\rho\,(\vec{u}\times\vec{\Omega})}_{\text{Coriolis force}} + \underbrace{\eta\nabla^2\vec{u}}_{\text{molecular friction}} - \underbrace{f_g}_{\text{gravity}}\,, \tag{3.1}$$

where $\vec{\Omega}$ is the rotational speed of the earth, ρ the air density and η the dynamic viscosity. The variables are the three-dimensional velocity vector $\vec{u}$ and the atmospheric pressure p.

Hence, there are three equations for four variables. The necessary fourth equation is derived from the conservation of mass and the approximation that the density of air is constant. It is called continuity equation and is given by

$$\nabla \cdot \vec{u} = 0 \,. \tag{3.2}$$

This incompressibility condition is well justified for atmospheric flow [2]. The two equations cover all flow situations except for thermal effects and completely describe flow regimes on all scales.

The nonlinearity, $(\vec{u} \cdot \nabla)\vec{u}$, in the Navier–Stokes equation (3.1) allows for a very rich and interesting variety of possible flow states and is, in particular, responsible for the existence of chaotic behaviour under certain flow conditions. Unfortunately, this non-linear structure does not permit to find general analytic solutions. Hence, one dilemma in describing the dynamics of the atmosphere is that though the exact equations of motions are known, it is in most cases not possible to solve them. In addition, real atmospheric flow states are mostly turbulent, which means that the fluid motion is very irregular and the variables are highly fluctuating; this makes it even more difficult to calculate the flow.

In order to focus on the mean flow and eliminate the details of the fluctuating motion of the fluid particles, the equations of motion are simplified. This approach is called Reynolds averaging and will be presented in the next section.

Reynolds-averaged equations are the basis for both analytic studies and numerical modelling of atmospheric flow. With the advent of powerful computers, numerical simulations are broadly used to describe and understand the behaviour of the atmosphere. This sounds easier, as it actually is, as the discretisation of the equations of motion and the finite accuracy of the numerical representation put some constraints on finding physically reasonable solutions. Nevertheless, computational fluid dynamics provides powerful tools that allow for new insights into the general behaviour of atmospheric flows and, of course, is the foundation of numerical weather prediction systems. For a general introduction to computational fluid dynamics, see, e.g., Ferziger et al. [25] and Fletcher [26]; the numerical modelling of meteorological phenomena is, e.g., described in [89].

Of course, notwithstanding the possibility of numerical simulation, Reynolds averaging allows for an analytic study of atmospheric flow problems leading to the concepts of boundary-layer meteorology, as shown in the following sections.

3.2.2 Reynolds-Averaged Equations

The variables in the Navier–Stokes equation (3.1) together with the continuity (3.2) describe the dynamics of the instantaneous variables which are fluctuating in a turbulent atmospheric flow. The exact details of the fluctuations are in most cases not relevant. What is important is the "macroscopic" effect of these fluctuations on the average behaviour of the flow. Hence, following an approach by Reynolds, the Navier–Stokes equation is rewritten to describe mean variables instead of instantaneous variables but at the same time covering the properties of the turbulent motion of the flow.

The mean variables are supposed to describe the average flow situation, i.e. the mean over all possible realisations of the turbulent flow. Thus, strictly speaking, this averaging process is an ensemble average. However, for practical purposes the average has to be taken either as time or space average which can only approximately represent the ensemble average.

Here, the temporal average is used and all variables in (3.1) and (3.2) are decomposed in a mean value averaged over a certain time interval and a fluctuating part, e.g. a wind speed component is written as

$$u_i(t) = U_i(t) + u'_i(t) \quad \text{with} \quad U_i(t) = \overline{u_i}(t) \ , \tag{3.3}$$

where the overbar denotes the temporal running average over an interval of length T:

$$\overline{u_i}(t) = \frac{1}{T} \int_t^{t+T} u_i(t')\, dt' . \tag{3.4}$$

The length of the time interval is arbitrary but fixed. It has to be chosen such that the time scale of turbulent fluctuations is separated from the time scale of the mean fluid motion. For practical purposes T is often set to 10 min.

With the definition (3.4) the average over the fluctuating part vanishes, i.e.

$$\overline{u'_i(t)} = 0 \ . \tag{3.5}$$

Now the variables $\vec{u}$ and p in (3.1) and (3.2) are substituted by their decompositions analogue to (3.3) and are then averaged. The result is that the Reynold-averaged equations for the mean value are almost identical to the original equations but with one additional term in the Navier–Stokes equation which describes an additional force on the mean flow due to turbulence. The Reynolds-averaged equations are given by

$$\rho \left(\partial_t \vec{U} + (\vec{U} \cdot \nabla)\, \vec{U} \right) = -\nabla p + 2\, \rho (\vec{U} \times \vec{\omega}) + \nu \nabla^2 \vec{U} - f_g - \underbrace{\nabla \tau_t}_{\text{turbulent momentum flux}} . \tag{3.6}$$

The essential new term is the tensor τ_t, which is in analogy to the viscous stress called turbulent stress tensor. It contains contributions

$$\tau_{t,ij} = \rho\, \overline{u'_i u'_j} \ , \tag{3.7}$$

which are due to the non-linearity in the Navier–Stokes equation and describe the a priori unknown correlations between the fluctuations of wind components. Note that

τ_t represents the influence of turbulent fluctuations on the average flow. Hence, in contrast to the viscous stress, which is basically a fluid property, the turbulent stress is a pure flow property.

Now what is the advantage of the Reynolds-averaged equations? At first sight the problem has not become simpler, in particular because the additional unknown contribution τ_t is introduced. The important point is that the averaged equations refer to a well-behaved mean flow without irregular fluctuations that seriously disturb the discussion of the various terms in the Navier–Stokes equation (3.1). Hence, the Reynolds-averaged (3.6) enables the identification of different flow regimes in the atmosphere according to the dominating physical effects. Of course, in order to find solutions for the averaged equations the effect of turbulence has to be described. There is no rigorous way to find a so-called turbulence closure, i.e. an additional equation for τ_t. Thus, a reasonable parameterisation which relates τ_t to the mean flow field has to be introduced. This will be discussed in Sect. 3.3.1.

3.3 Physical Concepts of Boundary-Layer Flow

3.3.1 Eddy Viscosity and Mixing Length

As already pointed out in Sect. 3.2.2, the turbulent stress tensor τ_t that describes the influence of turbulence on the mean velocity field has to be specified in order to find solutions of the Reynolds-averaged equations (3.6). The aim is to find additional equations for the unknown velocity fluctuations $\tau_{t,ij} = \rho\,\overline{u_i' u_j'}$ such that the problem is mathematically well defined.

There are different ways of finding such a turbulence closure which vary in their basic assumptions and their degree of complexity. In this book the focus will be on one particular approach based on the concepts of eddy viscosity and mixing length theory which relate the turbulent fluctuations to the mean flow. This process will be called parametrisation. A discussion of other closure schemes is beyond the scope of this book and can be found, e.g., in [2].

A straightforward way to relate the turbulent stress tensor to the flow is by assuming that the turbulent fluctuations are proportional to the shear of the mean velocity field. Thus,

$$\tau_{t,ij} = \rho\, K\, \partial_j U_i \; , \tag{3.8}$$

where K is the exchange coefficient of momentum. K is called eddy viscosity as this approach is analogous to the concept of parameterising molecular viscosity in a Newtonian fluid where K corresponds to the kinematic viscosity ν. The term “eddy” refers to the very qualitative idea that turbulent fluctuations can be imagined as vortices on different scales imposed on the mean flow motion. However, the difference between the two is that the eddy viscosity is related to the flow, while the kinematic viscosity is a property of the fluid only. Hence, K is not necessarily constant and can vary in different parts of the flow field.

In free turbulent flows such as free stream jets where boundaries can be neglected, the eddy viscosity K can successfully be considered as constant. This is no longer the case in the boundary layer of the lower atmosphere, where the presence of the ground has to be taken into consideration. Consequently, K has to be parametrised as well.

For this purpose Prandtl [96] suggested that fluid elements in turbulent flow can travel a certain distance before mixing with the surrounding flow. Hence, they can carry momentum (or other physical properties such as temperature or density) over a distance l, called the mixing length, without being perturbed. The interaction of these fluid elements with the new flow environment leads to fluctuations in the flow properties at this point.

This is illustrated for the vertical fluctuations of horizontal momentum in a stationary, horizontally homogeneous boundary flow where the velocity of the flow increases with height, as shown in Fig. 3.2. The assumption is that velocity fluctuations occur due to the vertical movements of fluid elements. For the sake of simplification the components of the velocity vector are $\vec{U} = (U, V, W)$ (analogous for the fluctuations) and the coordinate system is chosen such that $V = 0$. Moreover, the only relevant element of the turbulent stress tensor will be $\tau_{t,xz}$.

Consider the level z in Fig. 3.2 which has a mean speed of $U(z)$. If a fluid element moves downwards from a level $z+l$ with higher speed, it causes a velocity fluctuation

$$u' = U(z + l) - U(z) \approx l\,\partial_z U(z) \,, \tag{3.9}$$

where the mixing length l is sufficiently small to Taylor-expand the velocity profile. Since $U(z + l) \geq U(z)$, the gradient $\partial_z U(z)$ is positive. As a downward motion is related to a negative vertical velocity fluctuation w', one obtains

$$u'w' < 0 \,. \tag{3.10}$$

Because fluid elements arriving from $z - l$ lead to negative velocity fluctuations u', with w' being positive, (3.10) generally holds for velocity profiles with positive vertical gradient. Hence, to put it the other way round, if the flow velocity decreases towards the boundary, there is a net transport of horizontal momentum, i.e. $\rho\, u'w' < 0$, towards the boundary due to turbulence.

Equation (3.9) gives an estimation of the velocity fluctuations u' in terms of a yet unknown parameter l, the mixing length, which will be determined shortly. In order to estimate the vertical fluctuations w', Prandtl suggested to assume that they are of the same order of magnitude as u' but with an inverse sign because of the above considerations leading to (3.10); hence, $w' \approx -u'$. With (3.9) and the definition of the turbulent stress tensor in (3.7), this leads to

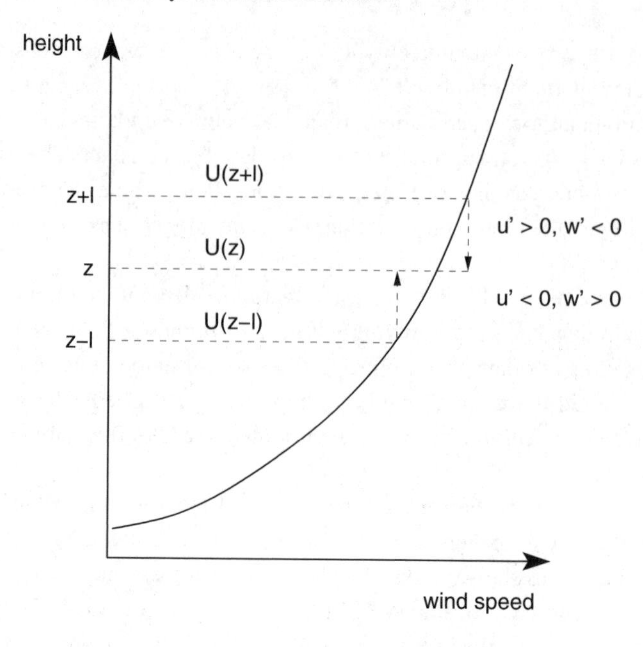

Fig. 3.2. Schematic illustration of the mixing length concept in terms of a wind profile. Fluctuations u' in horizontal velocity occur due to fluctuations w' of the vertical velocity of fluid elements over distance l. If the gradient of the mean wind speed U decreases with height, the product $u'w'$ is negative, which is equivalent to a net downward transport of horizontal momentum

$$\tau_{t,xz} = \rho\, \overline{u'w'} = -\rho\, l^2 (\partial_z U)^2 \,. \tag{3.11}$$

Note that all quantities except ρ can generally depend on the height z. Obviously, this equation is a strong simplification as the fluctuations of the different velocity components are treated separately in a very qualitative way without referring to any details of the turbulent statistics that might be behind the product $\overline{u'w'}$. However, the approach is justified as it leads to reasonable analytic results.

The above relation contains the important message that in contrast to the original assumption in (3.8), $\tau_{t,xy}$ depends on the square of the vertical velocity gradient or, in order to stay in the formalism, the eddy viscosity is given by

$$K = l^2\; |\partial_z U| \,. \tag{3.12}$$

As a last step, the unknown mixing length l has to be parameterised. For this purpose it is helpful to understand l as a length scale that puts a constraint on the typical eddy sizes. Hence, it is rather straightforward to assume that the size of l is directly related to the distance to the ground and, therefore, depends linearly on the height z. Hence,

$$l = \kappa z , \tag{3.13}$$

where κ is the von Karman constant, with $\kappa = 0.4$, which was determined empirically.

Finally, one obtains the parameterisation

$$\tau_{t,xz} = -\rho (\kappa z)^2 (\partial_z U)^2 . \tag{3.14}$$

This equation relates the relevant elements of the turbulent stress tensor, which refer to the transport of horizontal momentum in vertical direction, to the gradient of the mean velocity field and the geometry of the boundary flow. This relation is one way to describe the contribution of shear-generated turbulence within the planetary boundary layer. If plugged into the Reynolds-averaged equations (3.6), it can be used to derive analytic expressions for the vertical wind profiles in both the Ekman layer and the Prandtl layer.

3.3.2 Logarithmic Wind Speed Profile

With the considerations of the last section, the vertical wind profile in the surface layer can be derived. Again the flow field is stationary and horizontally homogeneous. Moreover, only neutral situations are considered, i.e. thermal effects, in particular, buoyancy, are not included so that wind shear is the main source of turbulence.

The basic assumption for the flow in the surface layer is that the turbulent momentum flux, $\tau_{t,xz}$, is independent of height. Though $\tau_{t,xz}$ is rather large in absolute values, because of the vicinity of the ground, it is supposed to vary only little within about the first 100 m above the surface. Thus, re-arranging (3.14) gives

$$\partial_z U = \sqrt{\frac{-\tau_{t,xz}}{\rho}} \frac{1}{\kappa z} = \frac{u_*}{\kappa z} , \tag{3.15}$$

where $u_* = \sqrt{-\tau_{t,xz}/\rho}$ at $z = 0$ is the so-called friction velocity which is constant with height. Hence, u_* contains information about the turbulent momentum transport near the ground for a certain flow situation.

The integration of (3.15) leads to the well-known logarithmic wind speed profile

$$U(z) = \frac{u_*}{\kappa} \ln \left(\frac{z}{z_0} \right) . \tag{3.16}$$

The integration constant z_0 is called roughness length as it is related to the surface roughness and varies over several orders of magnitude depending on the terrain type. Note that according to the above derivation, z_0 does not have to be the same for different u_*.

The logarithmic profile does not describe the instantaneous vertical wind speed at any time but a time-averaged wind profile not resolving the details of turbulent fluctuations in the flow. Hence, in order to compare (3.16) to measurements, the data have to be averaged over suitable time intervals of the order of 10 min. The log profile has been confirmed by many observations of wind speeds in near-neutral conditions. However, in situations where the atmospheric stratification is not neutral, the vertical wind profile has to be modified. This will be discussed in the next section.

3.4 Influence of Thermal Stratification on the Wind Profile

In the neutral logarithmic wind profile, thermal effects are not considered. However, in real situations, thermal effects play an important role and can strongly influence the wind profile.

During the course of the day the sun heats the ground, which in turn starts heating the air from below, and warmer air might rise due to buoyancy. If the upward movement of an air parcel is enhanced after it left its equilibrium position, the state of the atmosphere is called unstable. In contrast to this, the atmosphere is stable if the upward movement is dampened and the air parcel is driven back to the equilibrium position. This is primarily the case if the ground cools during clear nights. Onshore, the relatively small heat capacity of the ground allows for rather fast changes in the atmospheric stability, being strongly dependent on the irradiation of the sun, which, for example, leads to a considerable variation in the mean wind speeds at different heights over the day. This so-called diurnal cycle is due to changes in the thermal stratification of the atmosphere and is less pronounced over the ocean compared with land as the heat capacity of water is by far larger than that of soil.

Thermal stratification refers to the vertical temperature distribution in the atmosphere. Temperature gradients have a strong impact on the wind profile as they are the driving force of buoyancy, which is an additional source of turbulence in the flow. In contrast to mechanical turbulence generated by shear, the thermal turbulence is not inherently given by the current state of the flow.

In order to take the thermal stratification of the atmosphere into account, the logarithmic wind profile has to be modified. For this purpose Monin and Obukhov [76] suggested general correction functions Φ_m generalising the gradient approach that was used in (3.15) to derive the neutral profile. Hence,

$$\partial_z U = \frac{u_*}{\kappa z} \Phi_m(H, u_*, z) , \tag{3.17}$$

where Φ_m is a dimensionless function depending on the heat flux H, the friction velocity u_* and the height z. For the neutral case $\Phi_m(H, u_*, z)$ is clearly 1, while for other types of thermal stratification adequate functions have to be derived. In the following, Φ_m will be determined in the framework of the Monin–Obukhov theory.

3.4.1 Description of Thermal Stratification

When air parcels rise adiabatically, they expand due to the decreasing atmospheric pressure. As the work for the expansion is taken from the inner energy of the fluid element, its temperature decreases in this process. In neutral conditions the temperature of the new environment corresponds to that of the air parcel after it has moved, and there are no forces driving it further upwards or downwards. As these situations should be related to a vanishing temperature gradient, the potential temperature is used, which compensates for the cooling due to adiabatic expansion. In a simplified version, when the pressure is assumed to decrease linearly with height, the potential temperature, Θ, at a height z can be written as

$$\Theta(z) = T(z) + \Gamma\, z\,, \tag{3.18}$$

where T is the absolute temperature and Γ the so-called dry adiabatic temperature gradient given as

$$-\partial_z T = \Gamma \simeq 0.01\,\frac{K}{m}\,. \tag{3.19}$$

The vertical gradient of the potential temperature can now be used to distinguish different types of thermal stratification:

- $\partial_z \Theta = 0$: *neutral stratification*. The vertical temperature stratification does not influence the vertical momentum flux.
- $\partial_z \Theta < 0$: *unstable stratification*. The vertical temperature stratification enhances the vertical momentum transport due to buoyancy, e.g. on sunny days with high irradiation.
- $\partial_z \Theta > 0$: *stable stratification*. The vertical temperature stratification dampens the vertical momentum transport. This situtation mainly occurs during nights with clear sky.

The sign of the gradient only determines whether vertical movements of air parcels are enhanced or dampened. To quantitatively assess the degree of the atmospheric stability, the strength of the two different sources of atmospheric turbulence are compared. Therefore, the ratio between the power density of thermally driven turbulence, ϵ_A, and that of mechanical turbulence, ϵ_S, is considered and defines the Richardson number

$$Ri = -\frac{\epsilon_A}{\epsilon_S}\,. \tag{3.20}$$

The two contributions ϵ_A and ϵ_S cannot be measured directly. Hence, to take advantage of the Richardson number, several steps are necessary to estimate ϵ_A and ϵ_S by known quantities derived from thermodynamical and fluid dynamical considerations of the atmospheric flow. This will lead to the introduction of the Monin–Obukhov length as a typical scale to describe the influence of thermal stratification on the wind profile. Moreover, a robust method to extract the required quantities from measurement has to be given.

Thermally driven turbulence is supposed to be caused by buoyancy due to temperature fluctuations. Thus, the power density, ϵ_A, is given by

$$\epsilon_A = \frac{g\,\rho\,\overline{\Theta' w'}}{\Theta} . \tag{3.21}$$

Under the assumption that analogous to $\overline{u'w'}$, the turbulent transport $\overline{\Theta' w'}$ is also constant in the lower part of the atmosphere, this can be rewritten by introducing the vertical heat flux at the surface

$$H = c_p \rho\,\overline{\Theta' w'} \quad \text{at} \quad z = 0 , \tag{3.22}$$

and one obtains

$$\epsilon_A = \frac{g\,H}{c_p\,\Theta} , \tag{3.23}$$

where c_p is the heat capacity of the air.

On the other hand, the power density of the turbulence generated by shear is again estimated by the vertical velocity gradient

$$\epsilon_S = \rho\,u_*^2\,\partial_z U . \tag{3.24}$$

Using the generalised gradient approach in (3.17) ϵ_S is given by

$$\epsilon_S = \frac{\rho\,u_*^3}{\kappa\,z}\,\Phi_m(H, u_*, z) . \tag{3.25}$$

Note that the power density of the buoyancy-related turbulence ϵ_A is independent of height, while its shear-generated counterpart decreases with increasing height.

Finally, with (3.23) and (3.25) the Richardson number is expressed as

$$Ri = -\frac{\epsilon_A}{\epsilon_S} = -\frac{\kappa\,g\,H}{u_*^3\,\Theta\,c_p\,\rho}\,\frac{z}{\Phi_m(H, u_*, z)} . \tag{3.26}$$

In terms of extracting Ri from atmospheric measurements, this expression is still not very helpful, as the heat flux and the yet unknown functions Φ_m cannot directly be measured. However, the combination of variables in (3.26) contains the important length scale

$$L = -\frac{u_*^3\,\Theta\,c_p\,\rho}{\kappa\,g\,H} , \tag{3.27}$$

known as the Monin–Obukhov length. According to (3.26) the Richardson number Ri and L are, of course, closely related and can both serve as stability parameters to estimate the thermal stratification of the atmosphere.

The range of possible values of L is obviously $-\infty$ to ∞. The sign of L is determined by the sign of the heat flux H. Hence, L is positive in situations where the surface cools the air from below and negative if the surface heats the air. Concerning the physical meaning of the Monin–Obukhov length, Arya [2] points out that "in magnitude $|L|$ represents the thickness of the layer of dynamic influence near the surface in which shear or friction effects are always important". Hence, shear effects dominate for $z \ll |L|$ and buoyancy effects dominate for $z \gg |L|$. Thus, the scaled height z/L determines the relative importance of the two effects. Of course, one has to keep in mind that the theory only applies to the surface layer which extends only a few hundred metres from the ground.

3.4.2 Corrections to the Logarithmic Profile

With the newly introduced Monin–Obukhov length, it is clear that L summarizes the shear and buoyancy effects represented by u_* and H in an appropriate length scale. Thus, the thermal correction functions of the logarithmic profile are functions of z/L, i.e.

$$\Phi_m(H, u_*, z) = \Phi_m\left(\frac{z}{L}\right) . \tag{3.28}$$

The function Φ_m can only be determined empirically from experiments. The first formulations of $\Phi_m(z/L)$ were found independently by Businger [14] in 1966 and Dyer [19] in 1974. They have been confirmed by Webb [119], Hicks [42] and others. The functions are

$$\Phi_m\left(\frac{z}{L}\right) = \begin{cases} \frac{1}{(1-16\frac{z}{L})} & \text{for unstable stratification, } \frac{z}{L} < 0 \\ 1 & \text{for neutral stratification, } \frac{z}{L} = 0 \\ (1+5\frac{z}{L}) & \text{for stable stratification, } \frac{z}{L} > 0 . \end{cases} \tag{3.29}$$

Poulsen [95] and Panofsky [88] integrated the general velocity gradient in (3.17) using Φ_m from (3.29) and found

$$u(z) = \frac{u_*}{\kappa}\left(\ln\left(\frac{z}{z_0}\right) - \psi_m\left(\frac{z}{L}\right)\right) , \tag{3.30}$$

where

$$\psi_m\left(\frac{z}{L}\right) = \begin{cases} 2\ln\left(\frac{1+x}{2}\right) + \ln\left(\frac{1+x^2}{2}\right) - 2\,tan^{-1}(x) + \frac{\pi}{2} & \text{for } \frac{z}{L} < 0 \\ 0 & \text{for } \frac{z}{L} = 0 \\ -5\,\frac{z}{L} & \text{for } \frac{z}{L} > 0 , \end{cases} \tag{3.31}$$

with

$$x = \left(1 - 16\frac{z}{L}\right)^{1/4} . \tag{3.32}$$

It was shown by Holtslag [44] and Högström [43] that the functions ψ_m for stable situations are only valid in the range $0 < z/L < 0.5$. Therefore, Beljaars and Holtslag suggested a modified function for this case that is valid for larger heights up to $z/L \approx 7$. This approach is generally accepted and given by

$$\psi_m\left(\frac{z}{L}\right) = \frac{a\,z}{L} + b\left(\frac{z}{L} - \frac{c}{d}\right)\exp\left(-\frac{d\,z}{L}\right) + \frac{b\,c}{d}\,, \tag{3.33}$$

with $a = 1, b = 2/3, c = 5$ and $d = 0.35$.

Now the wind profile in the surface layer is given analytically for neutral and non-neutral stratification of the atmosphere. The following section comes back to the geostrophic wind, which is the driving force of the wind in the surface layer. The geostrophic wind is quantitatively related to the synoptic pressure systems, and the coupling between the geostrophic wind and the surface wind given by the friction velocity is described.

Finally, the vertical profile of the potential temperature can be derived analogous to the wind speed profile by using a dimensionless gradient Φ_h.

$$\Theta(z) = 0.74\,\frac{\Theta_*}{\kappa}\left(\ln\left(\frac{z}{z_0}\right) - \psi_h\left(\frac{z}{L}\right)\right) + \Theta_0, \quad \Theta_0 = \Theta(z_0)\,, \tag{3.34}$$

where

$$\psi_h\left(\frac{z}{L}\right) = \begin{cases} 2\,\ln\left(\frac{1+y}{2}\right);\, y = \frac{0.74}{\Phi_h\left(\frac{z}{L}\right)} & \text{for } \frac{z}{L} < 0 \\ 0 & \text{for } \frac{z}{L} = 0 \\ -\,6.4\frac{z}{L} & \text{for } \frac{z}{L} > 0 \end{cases} \tag{3.35}$$

and

$$\Phi_h\left(\frac{z}{L}\right) = 0.74\left(1 - \frac{9\,z}{L}\right)^{-1/2} \quad \text{for } \frac{z}{L} < 0\,. \tag{3.36}$$

Θ_* is a scaling factor similar to the friction velocity u_*.

3.4.3 Geostrophic Wind as Driving Force

In the mid-latitudes of the earth the wind fields are driven by large synoptic pressure systems, where the wind at higher altitudes that is not influenced by the friction of the ground is called geostrophic wind. It is characterised by a balance between pressured gradient force and Coriolis force such that the wind direction is parallel to the isobars. Hence, for a stationary and horizontal flow under the assumption that gravity, molecular friction and turbulent effects can be neglected in (3.6), the geostrophic wind, $\vec{G} = (u_g, v_g, 0)$, is given by

$$\nabla p = 2\,\rho(\vec{G} \times \vec{\omega})\,. \tag{3.37}$$

The vertical component of the Coriolis force is commonly ignored, so that the horizontal components of the geostrophic wind are described by

$$u_g = \frac{1}{2\,\rho\,\omega\,\sin\varphi}\frac{\partial p}{\partial y} \quad \text{and} \quad v_g = -\frac{1}{2\,\rho\,\omega\,\sin\varphi}\frac{\partial p}{\partial x}\,, \tag{3.38}$$

where $\omega \sin\varphi$ is the local component of the rotational speed of the earth. Introducing the Coriolis parameter $f = 2\,\omega\,sin\varphi$ (for Germany, $f \simeq 10^{-4} s^{-1}$), which depends on the latitude, the geostrophic wind, $\vec{G}$, can finally be written as

$$\vec{G} = \frac{1}{f\rho}(\nabla p \times \vec{k})\,, \tag{3.39}$$

with $\vec{k} = (0, 0, 1)$.

The coupling between geostrophic wind and the surface wind in the Prandtl layer is described by the geostrophic drag law, which was mathematically derived by Blackadar and Tennekes [10] based on "singular perturbation method".

The geostrophic drag law connects the geostrophic forcing, $\vec{G}$, by the undisturbed wind field in higher altitudes to the characteristic quantities in the surface layer defined by the friction velocity u_* and the roughness length z_0, while u_* describes the downward transport of horizontal momentum due to turbulent effects:

$$|G| = \frac{u_*}{\kappa}\sqrt{\left(\ln\left(\frac{u_*}{f\,z_0}\right) - A\right)^2 + B^2} \tag{3.40}$$

where κ is the von Karman constant, which is empirically determined and has a value of around 0.4. The parameter f covers the influence of the Coriolis force. The empirical constants A and B depend on the thermal stratification. For the neutral case the European Wind Atlas [115] states that $A = 1.8$ and $B = 4.5$.

The angle between geostrophic wind and surface wind due to the Coriolis force and the decreasing influence of friction with height is given by

$$\sin\alpha = \frac{-B\,u_*}{\kappa\,|G|}\,. \tag{3.41}$$

3.5 Conclusion

What has now been achieved? Starting from general considerations concerning the dominating physical effects influencing the flow in different layers, the full equations of motion of the atmosphere have been simplified to derive analytic expressions which quantitatively describe the vertical wind profile in the surface layer. The main idea is that the geostrophic wind in heights of around 1000 m drives the wind field in the lower layers by a downward transport of horizontal momentum.

In this process, turbulent mixing sustained by velocity and temperature fluctuations constitutes a coupling mechanism that is far more effective than molecular viscosity. The two main sources of turbulence in the flow are shear on the one hand and buoyancy on the other. While shear is an inherent flow property that can be described using the local velocity gradient, buoyancy is generated by thermal effects imposed by external sources, such as the heating of the ground by the sun.

The derivation of the neutral wind profile involves the concepts of eddy viscosity and mixing length, leading to a parametrisation of the shear-generated turbulent momentum flux. Thermal effects are covered by correcting the neutral profile with general non-dimensional functions that are determined empirically. However, the considerable achievement in this context is the identification of the parameters Richardson number Ri and Monin–Obukhov length L, which allow for a quantitative assessment of the thermal stratification of the lower atmosphere.

The derivation of the analytic wind profile requires a number of assumptions that might not be fulfilled in real atmospheric flows. Nevertheless, measurements confirm the usefulness of these profiles in many cases. However, it has to pointed out that the vertical profiles should be regarded as approximations, and the involved constants as quasi-constants. Hence, detailed measurements can in certain cases show a very different behaviour—naturally, in situations that are not covered by the assumptions. This will be further discussed in Chap. 7.

4

Physical Wind Power Prediction Systems

4.1 Introduction

This chapter describes in detail how a prediction system based on a physical approach calculates the expected power output using the wind speed forecast of a numerical weather prediction (NWP) system. The description focusses on the system *Previento*, which has been developed at the University of Oldenburg.

4.2 Basic Scheme

Physical prediction systems, in contrast to statistical systems, explicitly model the phenomena of the boundary layer that influence the power output of a wind farm. Figure 4.1 shows the basic scheme of the prediction system *Previento*. The numerical wind speed prediction is typically given by the NWP system with a rather coarse spatial resolution, e.g. 7×7 km^2 in case of the *Lokalmodell* provided by the German Weather Service. Thus, the NWP output has to be refined taking into account the local conditions at the specific site, in particular orography and direction-dependent surface roughness. To calculate the wind speed at hub height the thermal stratification of the atmosphere is modelled in detail. Then the wind speed is transferred to power output by the power curve, where either the certified curve or a site-specific curve which has been determined at the location can be used. In a wind farm the shadowing effects among the wind turbines can lead to reductions in power output up to 20% such that these farm effects are also considered. As a result *Previento* provides the predicted power output of a single wind farm. If measurement data from the wind farms are available, a statistical correction of systematic forecast errors is applied using linear regression techniques.

In practical use, energy traders and TSOs require the combined power output of many spatially dispersed wind farms in a region instead of that of a single wind farm.

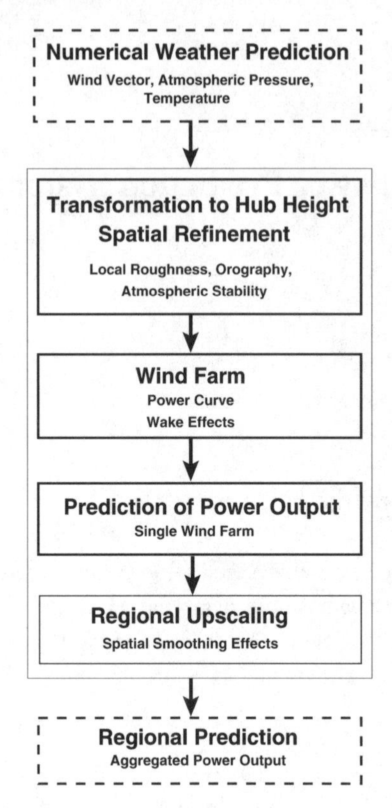

Fig. 4.1. Basic scheme of the wind power prediction system *Previento*, which is based on a physical approach. The input from the numerical weather prediction is locally refined and the wind speed at hub height is obtained with a physical model. The power prediction of a single wind farm is calculated by using the power curves of the wind turbines and considering the wake effects. The forecasts of representative wind farms in a region are scaled up under consideration of the spatial smoothing effects. The result is a prediction of the aggregated power output of all wind farms in a certain area

Consequently, *Previento* contains an advanced upscaling algorithm that determines the expected power output of all wind farms in a certain area based on a number of representative sites selected in an appropriate manner such that spatial smoothing effects are properly taken into account.

The most important modules of *Previento* are now described.

4.3 Detailed Roughness

The derivation of the logarithmic wind profile (Chap 3) suggests the concept of a typical length z_0 that describes the surface roughness. For onshore sites it is assumed to be independent of the wind field. Based on empirical data, different surface types have been classified and associated with a characteristic roughness length. While z_0 is relatively small, i.e. in the order of magnitude of 10^{-3} m, for smooth surfaces such as sand or water, it increases for grassland ($z_0 \approx 10^{-2}$ m) and urban areas ($z_0 \approx$ 1 m). A detailed description can be found, e.g. in the European Wind Atlas [113,115].

The predicted wind speed refers to the roughness length of the numerical prediction model, which will be denoted as $z_{0,\mathrm{NWP}}$ and represents the average roughness of the corresponding grid cell. Obviously, $z_{0,\mathrm{NWP}}$ might not be the optimal choice for the location of the wind farm. It is the task of the spatial refinement to replace the given roughness length $z_{0,\mathrm{NWP}}$ by a more adequate value $z_{0,\mathrm{local}}$ which reflects the on-site conditions.

In order to transform the wind speed u_{NWP} given with respect to a certain roughness length $z_{0,\mathrm{NWP}}$ to the local roughness length $z_{0,\mathrm{local}}$, the geostrophic drag law (3.40) is used twice. First, it is used "upwards" to calculate the geostrophic wind G based on $u_{*,\mathrm{NWP}}$ and $z_{0,\mathrm{NWP}}$. This geostrophic wind does not depend on surface effects and, therefore, can directly be used "downwards" to calculate to obtain u_{local}—a step that, unfortunately, requires to solve for $u_{*,\mathrm{local}}$ in (3.40). As a result the vertical wind profile that refers to the local roughness conditions can be calculated using the logarithmic wind profile (3.16).

In addition, as the structure of the surface typically varies around the site, the roughness length has to be described depending on the distance from the site and the direction of the wind. For this purpose the area around the site is divided into 12 sectors having 30° each, and roughness changes in radial direction along the sectors are taken into consideration.

Downstream of a roughness change an internal boundary layer (IBL) develops due to the discontinuity of the surface. This is illustrated in Fig. 4.2 for a situation with two roughness changes. If the wind approaches from the left-hand side, the roughness change at distance x_1 generates an IBL. This leads to a vertical wind profile which is dominated by z_{01} above the IBL and by z_{02} below. Close to the site, another roughness change at x_2 causes a second IBL to develop such that the wind profile at the location of the wind turbine consists of three different parts which have to be calculated iteratively.

According to Panofsky [88], the height h of the IBL at a distance x_n downstream of a roughness change is given by

$$\frac{h}{z_{0\,\max}}\left(\ln\left(\frac{h}{z_{0\,\max}}\right)-1\right)=0.9\,\frac{x_n}{z_{0\,\max}}\,, \tag{4.1}$$

where

$$z_{0\,\max}=\max(z_{0n},z_{0(n+1)})\,. \tag{4.2}$$

z_{0n} and $z_{0(n+1)}$ refer to the roughness lengths upstream and downstream, respectively, of the roughness change.

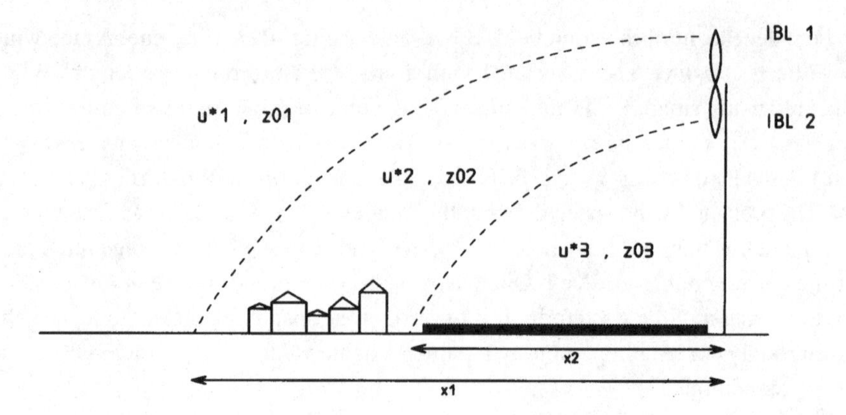

Fig. 4.2. Development of internal boundary layers (IBL) at two roughness changes x_1 and x_2. The characteristic quantities within the IBL are the friction velocity $u_{*,i}$ and the roughness length $z_{0,i}$

This model has been developed for short distances from the roughness change. For larger distances, Sempreviva et al. [102] suggested an empirical correction based on measurement data. This approach introduces additional heights at 9% and 30% of h. In case of two discontinuities in the surface roughness the wind profile at the site is then described by

$$u(z) = \begin{cases} \frac{u_{*1}}{\kappa} \ln(\frac{z}{z_{01}}) & \text{for } 0.3 \cdot h_1 \leq z \\ u'' + (u''' - u'')\frac{\ln(\frac{z}{0.3 \cdot h_2})}{\ln(\frac{0.3 \cdot h_1}{0.3 \cdot h_2})} & \text{for } 0.3 \cdot h_2 \leq z < 0.3 \cdot h_1 \\ u' + (u'' - u')\frac{\ln(\frac{z}{0.09 \cdot h_2})}{\ln(\frac{0.3 \cdot h_2}{0.09 \cdot h_2})} & \text{for } 0.09 \cdot h_2 \leq z < 0.3 \cdot h_2 \\ \frac{u_{*2}}{\kappa} \ln(\frac{z}{z_{03}}) & \text{for } z < 0.09 \cdot h_2 \end{cases} \tag{4.3}$$

with

$$u' = \frac{u_{*3}}{\kappa} \ln\left(\frac{0.09 \cdot h_2}{z_{03}}\right) , \tag{4.4}$$

$$u'' = \frac{u_{*2}}{\kappa} \ln\left(\frac{0.3 \cdot h_2}{z_{02}}\right) , \tag{4.5}$$

$$u''' = \frac{u_{*1}}{\kappa} \ln\left(\frac{0.3 \cdot h_1}{z_{01}}\right) . \tag{4.6}$$

h_1, h_2 and h_3 refer to the heights of the corresponding IBLs at the site. As the wind profile consists of different parts, the friction velocities u_{*n} cannot be determined by the geostrophic drag law such that they are approximated (see [115]) by the matching relation

$$u_{*2} = u_{*1} \frac{\ln(\frac{h_1}{z_{01}})}{\ln(\frac{h_2}{z_{02}})} . \tag{4.7}$$

The friction velocities are then calculated iteratively starting with the IBL farthest from the site.

With growing distance from the roughness change, the clear structure of the IBL decays. Due to this mixing, the influence of discontinuities of the surface roughness that are far away from the site is less pronounced compared with those close to the site. This is considered by an exponentially decreasing weight factor

$$\ln(z_{02}) = \ln(z_{03}) + W_n \ln\left(\frac{z_{02}}{z_{03}}\right) , \tag{4.8}$$

with

$$W_n = e^{\left(\frac{x_n}{10000m}\right)} . \tag{4.9}$$

Hence, the influence of roughness changes at a distance x_n is dampened such that the surface structure of the terrain which is more than 10 km away from the site does not contribute to the wind profile.

The procedure described above eliminates the influence of the original mesoscale roughness given by the NWP system and re-calculates the wind speed in terms of the local direction-dependent roughness.

4.4 Thermal Stratification

As discussed from a theoretical point of view in Chap. 3, the thermal stratification of the lower atmosphere strongly influences the coupling between flow layers at different heights and, therefore, alters the vertical wind speed profile. In Central Europe the atmosphere is on average slightly stable, leading to increased wind speeds at heights above 50 m compared to the logarithmic wind profile. For resource assessments, where annual averages of the wind conditions are described, a general correction of the logarithmic wind profile based on climatological means of the heat flux at the surface (see [113, 115]) is sufficient to account for thermal stratification. In contrast to this, for short-term predictions it is required that the wind profiles be individually corrected according to the thermal stratification of the prevailing situation.

The coupling between different flow layers due to turbulent mixing can vary significantly within hours. Figure 4.3 shows time series of measured wind speeds at 10 m and 80 m height together with the corresponding temperature difference between the two heights indicating thermal stratification. It can clearly be seen that in

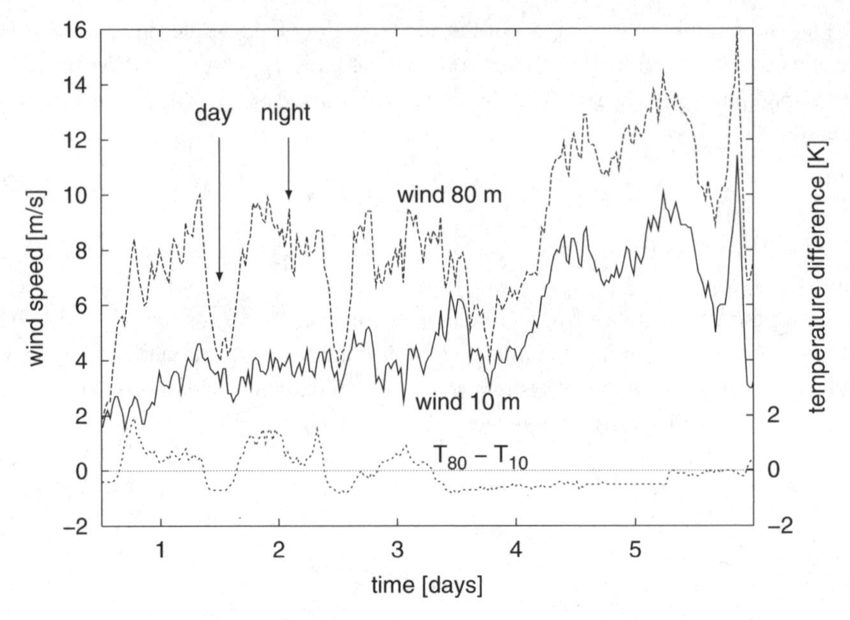

Fig. 4.3. Example time series of wind speeds at 10 m and 80 m. The difference between the wind speeds varies from 0.5 m/s in daytime to 5 m/s during the night. These pronounced variations are due to thermal stratification, which is indicated here by the temperature difference $(T_{80} - T_{10})$ between the two heights

the daytime, when the temperature near the ground (10 m) is greater than at 80 m, the difference between the wind speeds is small. This is due to solar irradiation, which heats the ground and causes buoyancy such that turbulent mixing leads to an effective coupling between the wind fields in the surface layer. During nighttime the temperature distribution changes sign because of the cooling of the ground. This inversion dampens turbulent mixing and, hence, decouples the wind speed at different heights, leading to pronounced differences between wind speeds.

As a result of this effect the mean values of the wind speed show a pronounced diurnal cycle, which is shown in Fig. 4.4. At 10 m-height the mean wind speed has a maximum at noon and a minimum around midnight. This behaviour changes with increasing height, so that at 200 m the diurnal cycle is inverse, with a broad minimum in daytime and maximum wind speeds at night. Hence, the better the coupling between the atmospheric layers during the day, the more horizontal momentum is transferred downwards from flow layers at large heights to those near the ground.

4.4.1 Parametrisation of the Monin–Obukhov Length

In order to obtain a thermally corrected wind profile in each situation, NWP systems typically offer the heat flux H at ground level, and this would be convenient with

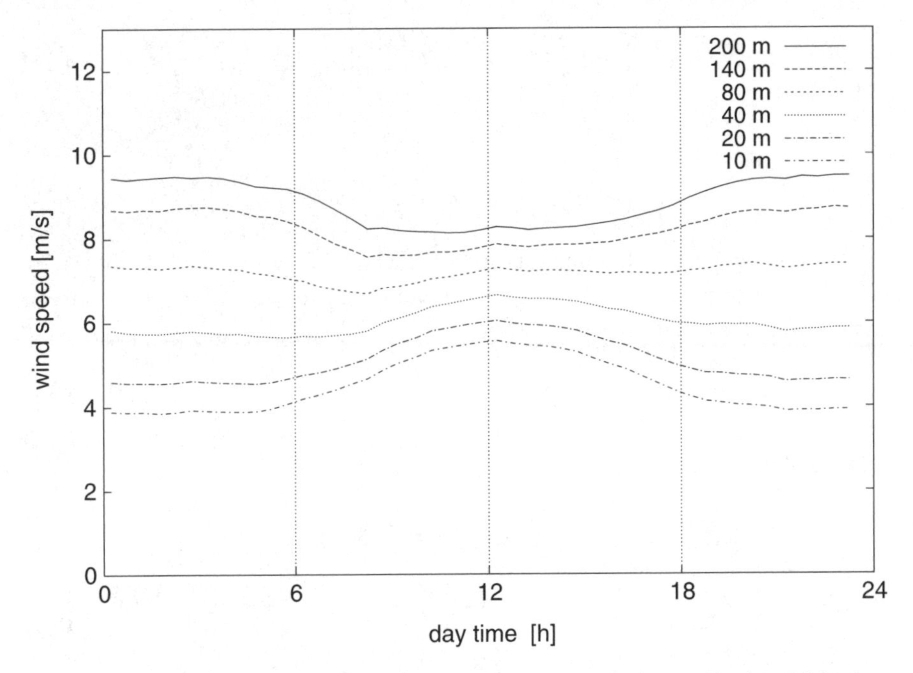

Fig. 4.4. Mean measured wind speeds at different heights dependent on the time of the day

regard to calculating the Monin–Obukhov length L according to (3.27). However, sonic anemometers which are needed to measure the heat flux directly are seldom available. Hence, for use in *Previento* an alternative way of parametrising L to determine the thermal stratification has been chosen which is based on temperature differences and wind speed differences between different heights. Among a number of analytic approaches the gradient method by deBruin [16] is the only one which does not require that wind speeds and temperatures be given at the same heights.

The main idea of the deBruins approach is to find an analytic approximation of the Monin–Obukhov length L in terms of measured bulk Richardson numbers. In Sect. 3.4 L has been introduced as a stability parameter expressing the ratio between mechanical turbulence generated by wind shear and thermally induced turbulence due to buoyancy. Alternatively, (3.27) L can also be written as

$$L = \frac{T}{g\kappa} \frac{u_*^2}{\Theta_*} , \tag{4.10}$$

where the sensible heat flux at ground level is given by $H = -\rho c_p u_* \Theta_*$ (c.f. (3.22) and (3.34)).

Let $h_{1\mathrm{T}}$, $h_{2\mathrm{T}}$ be the pair of heights to determine the temperature difference, $\Delta\Theta$, and $h_{1\mathrm{u}}$, $h_{2\mathrm{u}}$ be the pair to determine the wind speed difference, Δu. According to the vertical profiles of wind speed and temperature, these differences are given by

$$\Delta u = \frac{u_*}{\kappa}\left(\ln\left(\frac{h_{2\mathrm{U}}}{h_{1\mathrm{U}}}\right) - \psi_m\left(\frac{h_{2\mathrm{U}}}{L}\right) + \psi_m\left(\frac{h_{1\mathrm{U}}}{L}\right)\right) \tag{4.11}$$

and

$$\Delta\Theta = 0.74\,\frac{\Theta_*}{\kappa}\left(\ln\left(\frac{h_{2\mathrm{T}}}{h_{1\mathrm{T}}}\right) - \psi_h\left(\frac{h_{2\mathrm{T}}}{L}\right) + \psi_h\left(\frac{h_{1\mathrm{T}}}{L}\right)\right) . \tag{4.12}$$

The task now is to determine L based on Δu and $\Delta\Theta$, which are either given by measurement or prediction. Solving (4.11) and (4.12) for u_* and Θ_*, respectively, and inserting the result in (4.10) leads to

$$L = a\,L_0 \frac{f_h}{f_m^2} , \tag{4.13}$$

with the abbreviations

$$L_0 := \frac{T}{g}\,\frac{\Delta u^2}{\Delta\Theta} \tag{4.14}$$

and

$$a := \frac{\ln\left(\frac{h_{1\mathrm{T}}}{h_{2\mathrm{T}}}\right)}{\left[\ln\left(\frac{h_{1\mathrm{U}}}{h_{2\mathrm{U}}}\right)\right]^2} . \tag{4.15}$$

The functions f_h and f_m are defined by

$$f_h := \left[1 - \frac{\psi_h\left(\frac{h_{1\mathrm{T}}}{L}\right) - \psi_h\left(\frac{h_{2\mathrm{T}}}{L}\right)}{\ln(h_{1\mathrm{T}}) - \ln(h_{2\mathrm{T}})}\right] \quad \text{and} \quad f_m := \left[1 - \frac{\psi_m\left(\frac{h_{1\mathrm{U}}}{L}\right) - \psi_m\left(\frac{h_{2\mathrm{U}}}{L}\right)}{\ln(h_{1\mathrm{U}}) - \ln(h_{2\mathrm{U}})}\right] . \tag{4.16}$$

The two functions f_h and f_m still contain the unknown parameter L. In order to replace L by quantities that are readily available, bulk Richardson numbers are introduced. Due to the fact that wind speeds and temperatures are given on arbitrary heights, two different bulk Richardson numbers are defined:

$$\mathrm{Ri}_{\mathrm{bT}} = \frac{h_{1\mathrm{T}} - h_{2\mathrm{T}}}{L_0} \quad \text{and} \quad \mathrm{Ri}_{\mathrm{bU}} = \frac{h_{1\mathrm{U}} - h_{2\mathrm{U}}}{L_0} . \tag{4.17}$$

Inspired by the correction functions by Dyer as well as Beljaars and Holtslag (c.f. Sect. 3.4) used to correct the logarithmic profiles , f_h and f_m are approximated as

$$f_h = (1 - a_{\mathrm{hu}}\,\mathrm{Ri}_{\mathrm{bT}})^{-1/2} \quad \text{and} \quad f_m = (1 - a_{\mathrm{mu}}\,\mathrm{Ri}_{\mathrm{bU}})^{-1/4} \tag{4.18}$$

for unstable conditions, i.e. where $\Delta\Theta < 0$. For stable situations the approximation is given by

$$f_h = (1 + a_{\mathrm{hs}}\,\mathrm{Ri}_{\mathrm{bT}}^{0.5}) \quad \text{and} \quad f_m^2 = (1 + a_{\mathrm{hs}}\,\mathrm{Ri}_{\mathrm{bU}}^{1.5}) . \tag{4.19}$$

The four coefficients a_{hu}, a_{mu},a_{hs} and a_{hs} have to be determined only once from measurements which will be explained later.

Finally, the analytic approximations derived in (4.18) and (4.19) are inserted into the definition of L in (4.13). For the unstable case this leads to

$$L = a\, L_0 \sqrt{\frac{1 - a_{\mathrm{mu}}\, \mathrm{Ri}_{\mathrm{bU}}}{1 - a_{\mathrm{hu}}\, \mathrm{Ri}_{\mathrm{bT}}}} , \tag{4.20}$$

and for stable situations one obtains

$$L = a\, L_0 \frac{1 + a_{\mathrm{hs}}\, \mathrm{Ri}_{\mathrm{bt}}^{0.5}}{1 + a_{\mathrm{ms}}\, \mathrm{Ri}_{\mathrm{bU}}^{1.5}} . \tag{4.21}$$

To calculate the constants a_{hu}, a_{mu}, a_{hs}, and a_{hs}, one combination of heat flux H and friction velocity u_* obtained from previous experience is sufficient (as discussed by deBruin et al. [16]). For the unstable case a suitable choice of these two parameters is $H = 100$ W/m^2 and $u_* = 0.2$ m/s. Using the annual mean temperature and (4.10), L can be determined, so also can be the exact values of f_h and f_m according to (4.16). Comparing them with the approximations (4.18) gives the coefficients a_{hu} and a_{mu}. For stable situations the relation $L = \sqrt{h_{1\mathrm{U}} h_{1\mathrm{T}}}$ is used together with (4.16) and the approximations (4.19) to determine a_{hs} and a_{ms}.

In Fig. 4.5 the wind profiles for different thermal stratifications are illustrated, where the wind speed at 10 m is always 5 m/s. Comparing a neutral profile, an unstable profile with a temperature difference $\Delta\Theta = -2$ K and a stable profile with a difference of $\Delta\Theta = 4$ K shows that it is necessary to consider thermal stratification in the extrapolation of the 10-m wind speed to greater heights.

4.5 Orography

Complex topographic structures of the Earth's surface such as hills or valleys profoundly change the flow and affect wind speed and direction. At the crest of hills the wind speed can be significantly increased due to a compression of streamlines. These speed-up effects are, of course, one of the reasons to erect wind turbines on top of hills. The presence of valleys typically lead to channeling effects, where the wind direction deviates considerably from the unperturbed air flow.

Turbulent flow in complex terrain is a broad subject where a lot of progress has been made over the last 50 years in terms of theoretical concepts as well as numerical modelling as described, e.g., by Wood [122]. It is beyond the scope of this book to give a comprehensive introduction to the description of flow in complex terrain. Interested readers are referred to text books on boundary-layer meteorology, e.g. Stull [107], Arya [2] or Lalas and Ratto [57].

There are several numerical models which simulate these effects based on the equations of motion (Chap. 5). However, comparison of results from mesoscale models, e.g. *Gesima* [106], *MM5* [24] or *Fitnah* [39], with measured data revealed rather

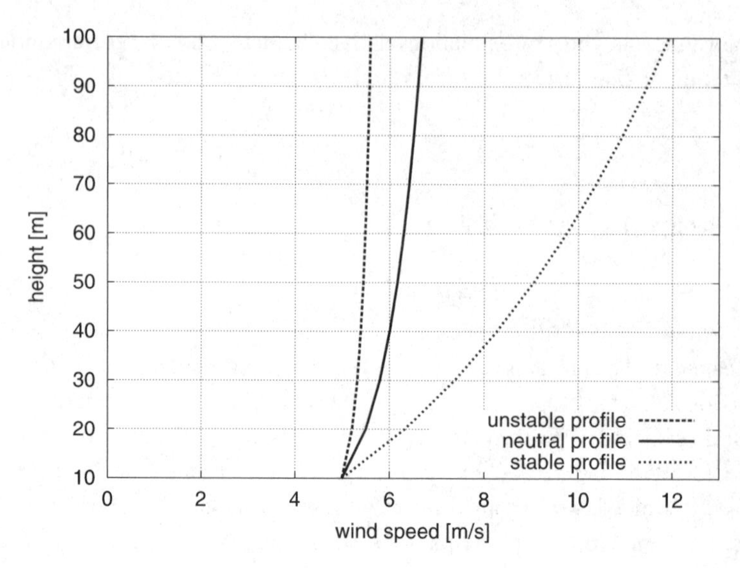

Fig. 4.5. Wind profiles for different types of thermal stratification having a wind speed of 5 m/s at 10-m height. The comparison shows the neutral profile (*solid line*) as well as a profile for an unstable situation with a temperature difference $\Delta\Theta = -2\ K$ between 10-m and 80-m height (*dashed line*) and a profile for a stable situation corresponding to a temperature difference of $\Delta\Theta = 4$ K (*dotted line*)

large deviations, showing that either approximations within the models or the input data used as initial conditions are not sufficient. In addition, the amount of data collection and handling that is necessary to operate the numerical models is relatively large and costly.

Previento accounts for effects of non-homogeneous terrain with the orographic model of the European Wind Atlas (EWA). This methodology is called BZ model (Bessel expansion on a zooming grid), a full description of which is given by Troen in [115] and [113]. The BZ model is a modification of a simple flow model suggested by Jackson and Hunt [47] (cf. [114]). It accounts for effects due to the undulations of the terrain on the scale of 10 km. The main advantage of the EWA model over the original Jackson and Hunt approach is its formulation in polar coordinates, which inherently leads to a high resolution at the origin where the site of the wind turbine is placed.

The Jackson and Hunt approach describes the flow in complex terrain as a small perturbation to the logarithmic profile of the unperturbed flow. Hence, the wind vector is given by

$$\vec{u} = \vec{u_0} + \vec{u}' = \vec{u_0} + (u', v', w') , \tag{4.22}$$

where $\vec{u_0}$ is the unperturbed velocity field and $\vec{u}' = (u', v', w')$ a perturbation. This approach is used to linearise the equations of motion (Chap. 3) around the flow state $\vec{u_0}$.

The orographic effects of terrain elevations on the unperturbed wind flow are calculated in two steps. First, frictional effects are neglected and the flow perturbation is determined with a potential flow. Second, the potential flow is corrected to include surface friction and turbulent momentum transport.

4.5.1 Potential Flow

The perturbation $\vec{u}' = (u', v', w')$ is assumed to be related to a potential field Φ such that

$$\vec{u'} = \nabla \Phi \, . \tag{4.23}$$

Note that potential flow is by definition irrotational and does not account for the viscosity or turbulence of the fluid. Due to the incompressibility of air the potential Φ yields the Laplace equation

$$\nabla^2 \Phi = 0 \, . \tag{4.24}$$

The geometry of the flow problem is that the perturbation is to be described from the point of view of a wind turbine which has a single point as horizontal coordinate and a vertical extension. This quite naturally suggests one to find solutions in cylindrical polar coordinates. It is shown in standard mathematical textbooks that the Laplace equation (4.24) is generally solved by a linear combination of the form

$$\Phi(r, \phi, z) = \sum_n \sum_j K_{nj} \Phi_{nj}(r, \phi, z) = \sum_n \sum_j K_{nj} \, J_n(\alpha_j r) \, e^{in\phi} e^{-\alpha_j z} \, , \tag{4.25}$$

where K_{nj} and α_j are coefficients to be determined by the boundary conditions. J_n is the nth order Bessel function, r the distance from the site, ϕ the polar angle and z the height above ground level.

The perturbation is assumed to vanish at a distance R from the site where the flow is not perturbed by the terrain elevation. Hence, the radial part of the solutions in (4.25) R must have a zero for $r = R$ and, hence,

$$\alpha_j = \frac{c_j^n}{R} \, , \tag{4.26}$$

where c_j^n is the jth zero of J_n. Note that this also provides each solution with a specific vertical length scale $L_j = R/c_j^1$, which is the characteristic extension of the perturbation.

Obtaining K_{nj} requires some more effort, but these coefficients will contain all the information concerning the terrain. The boundary condition at the surface height ($z = 0$) is given by the vertical speed w_0 that is induced by the terrain elevation $h(r, \phi)$ for an unperturbed wind field $\vec{u_0}$. Hence,

$$w_0(r,\phi) = \left.\frac{\partial \Phi}{\partial z}\right|_{z=0} = \vec{u_0} \cdot \nabla h(r,\phi) \ . \tag{4.27}$$

In terms of the solution (4.25) and the condition (4.26), this leads to

$$w_0(r,\phi) = \sum_n \sum_j \left(-\frac{c_j^n}{R}\right) K_{nj}\, J_n\!\left(c_j^n \frac{r}{R}\right) e^{in\phi} \ , \tag{4.28}$$

which is a Bessel series in r and a Fourier series in ϕ. Using the fact that the Bessel functions form an orthogonal basis of the radial function space while the Fourier modes provide an orthogonal basis of the angular functions, the coefficients K_{nj} are, as usual, determined by integration, and one obtains

$$K_{nj} = -\frac{1}{c_j^n R} \frac{2}{\pi \left[J_{n+1}(c_j^n)\right]^2} \int_0^R \int_0^{2\pi} w_0(r,\phi)\, J_n\!\left(c_j^n \frac{r}{R}\right) e^{-in\phi}\, r\, dr\, d\phi \ . \tag{4.29}$$

Thus, for a specific wind field $\vec{u_0}$ the influence of the terrain elevation $\nabla h(r,\phi)$ is expressed by a perturbation of the form (4.25), where the orographic information is implicitly given by coefficients K_{nj} according to (4.29).

In the Wind Atlas model the final result of the first model step is a series of coefficients K_{1j} for $n = 1$, which are determined numerically by integrating over a digitized topographical map. The site of interest is, of course, placed at the origin (i.e. $r = 0$), where the numerical grid has the highest resolution.

In a second step the effect of turbulent momentum transport is regarded by correcting the inviscid solution (4.25) at the location of the site ($r = 0$) with a logarithmic approach. The correction is important up to a height $l_j \ll L_j$, given by the length scale of the solution. In this model the value of l_j is calculated according to

$$l_j = 0.3 z_{0j} \left(\frac{L_j}{z_{0j}}\right)^{0.67} \ , \tag{4.30}$$

where z_{0j} is the roughness length that is relevant for the range of the vertical profile given by l_j (c.f. Sect. 4.3) with $z_{0j} = z_0$, for homogeneous terrain with no roughness changes in upwind direction.

The actual correction $\Delta\vec{u_j}$ to the potential flow for heights $z \leq l_j$ is then given by

$$\frac{\Delta\vec{u_j}(z)}{|\Delta\vec{u_0}(z)|} = \frac{|\Delta\vec{u_0}(L_j)|^2}{|\Delta\vec{u_0}(z)|^2} \nabla \Phi_j \ . \tag{4.31}$$

The result is a two-dimensional look-up table that provides a correction factor for each wind speed and each sector.

4.6 Farm Effects

Wind farms can typically comprise a considerable number of single wind turbines. Due to the short distances between the machines, shadowing effects occur where

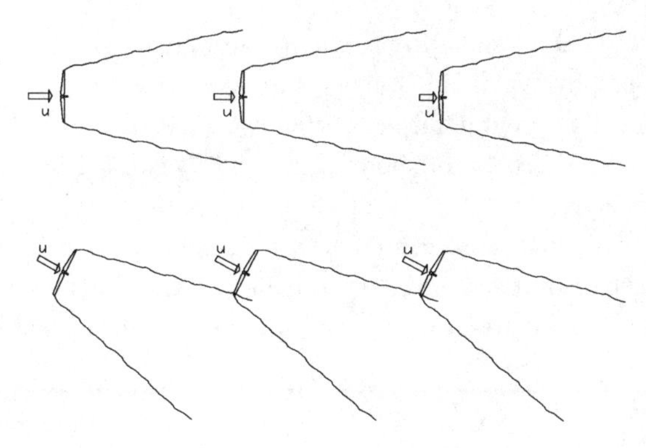

Fig. 4.6. Shadowing effect in a wind farm with three turbines in a row for different wind directions. For wind directions from the west (*top*), two machines are fully exposed to the wakes of the upstream machines, leading to a reduced farm efficiency. For north-westerly directions (bottom) the wake effect vanishes

turbines that are upstream may profoundly reduce the wind power that is available for the turbines downstream. In an area behind the rotor disc the wind speed is decreased due to the momentum that has been extracted from the flow to turn the rotor. This area is called wake and is illustrated in Fig. 4.6.

Waldl [118] has investigated wakes and their influence on the power production of wind farms in detail. He used wake models formulated by Ainslie [1], which are based on the dynamics of a axisymmetric flow field, and the simpler Risø model [48, 55], which showed a rather good performance [9]. Figure 4.7 shows the simulation of a wake by the Ainslie model compared to corresponding measurements of the wind

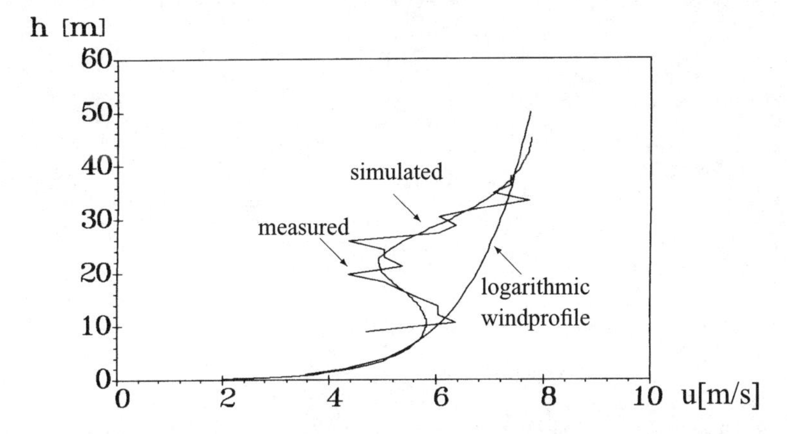

Fig. 4.7. Vertical wind profile of the measured and simulated wake effect at a distance of three rotor diameters downstream of a turbine. For comparison the unperturbed logarithmic profile is shown (from [118])

speed. For the prediction system *Previento* the farm layout programme *Flap*, which has been developed by Waldl [117], is used to take the wake effects inside a wind farm into account. This programme has currently been enhanced by Lange et al. [62] to include also wake effects for offshore wind farms by considering the effects of the marine boundary layer.

The reduction of wind speed behind the rotor compared to the free wind speed is quite profound and leads to a reduced performance of the wind farm for certain wind directions. This is quantitatively described by a direction-dependent farm efficiency factor defined by

$$\eta_{\text{farm}} = \frac{\sum_{\text{wec}} P_{\text{wec,farm}}}{\sum_{\text{wec}} P_{\text{wec,free}}}, \tag{4.32}$$

which is the ratio between the sum of the maximum power output of the turbines in a wind farm under consideration of the wake effects and the aggregated power of the free turbines.

Figure 4.8 illustrates η_{farm} according to (4.32) for a small wind farm which consists of two rows of turbines. The considerable decrease of the farm efficiency at

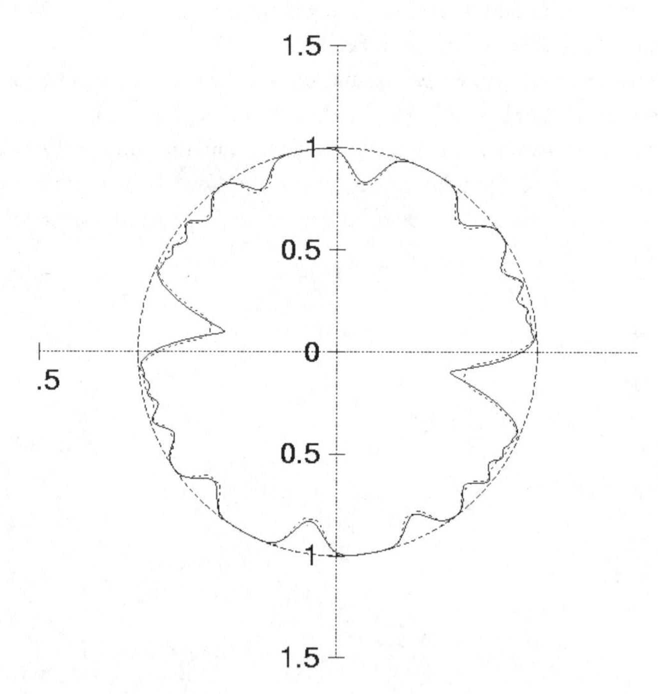

Fig. 4.8. Direction-dependent farm efficiency factor of a small wind farm which consists of two rows of turbines. "1" refers to situations where the power output of the wind farm is equal to the power output of the same number of unperturbed turbines. The farm efficiency decreases considerably if the rows are parallel to the wind vector (100° and 280°)

100° and 280° is due to the farm geometry, as in these cases the wind vector is parallel to the rows of turbines, leading to a maximum shadowing effect. For small wind farms the average farm efficiency is around 90%, while for large wind farms it can drop to 80%.

5

Data

Abstract. The data used for the investigations in this book is described in detail. Due to the fact that measurement data is provided as time-series of point-measurements while numerical weather prediction are based on computational grids this chapter depicts how the data have to be carefully pre-processed in order to compare and evaluate them.

5.1 Introduction

The data that are used in the investigations of the next chapters fall in two categories: prediction and measurement. Though it is quite straightforward to verify predictions at a certain site with the corresponding measured data, the origin and the way in which both data types are calculated or collected, respectively, deserve some discussion.

Predictions, on the one hand, are the result of a numerical integration of the equations of motions of the atmosphere discretised on a computational grid with a rather coarse resolution of the order of 10×10 km^2. Hence, the predicted variables represent a spatial average over the grid cell rather than a point value. This has to be kept in mind, in particular, with respect to the wind vector which is sensitive to the local conditions, while scalar quantities are not that much affected.

To evaluate the accuracy of these predictions it is desirable to compare them to the "real" values of the meteorological variables. As usual in physics, measurements, which are not arbitrarily accurate, are used to approximately assess reality. In contrast to the predictions, the measurements are taken at points. For example, the cup anemometers that are usually used to measure wind speeds have a typical diameter of the order of 0.1 m, and their rather short response time due to inertia enables a very localised measurement in space and time.

The common approach to compare both data types at a given location is time averaging the measured time series over a suitable time interval. This primarily aims at eliminating the high frequent fluctuations due to turbulence on the time scale below 10 min as these small scales cannot at all be resolved by numerical weather prediction (NWP). Experience shows that a reasonable averaging period is around half an hour to one hour. This time interval captures the temporal variations of the meteorological parameters at a synoptic scale and is, therefore, believed to correspond to the spatial average provided by the prediction values of the NWP systems.

5.2 Numerical Weather Predictions

All prediction data used in this work are provided by the NWP of the German Weather Service (DWD). The investigations are based on the results of the *Deutschlandmodell* (DM) and the *Lokalmodell* (LM). The DM version 4 [100] has a spatial resolution of 14×14 km^2 horizontally, and the LM 7×7 km^2 . In the vertical dimension the model comprises 30 levels extending up to 25 km height, where the two lowest of these levels are approximately at 34 m and 111 m height. The domain of the DM completely covers Central Europe and the British Isles. The boundary values of the DM are set by the European model EM with a resolution of about 55×55 km^2, which is itself nested into the global model GM that spans the whole earth. The DM data are received as points on the computational grid. To determine the forecast values at arbitrary locations the values at the four nearest grid points are interpolated using inverse distance weights.

From the two main runs of the DM and LM started every day at 00 UTC, (= GMT) and 12 UTC, only data from the earlier run are available at the prediction times +6, +12, +18, +24, +36 and +48 h. These times are counted relative to the starting time, 00 UTC, of the forecast run and, hence, directly correspond to the actual time of day in UTC.

Previous investigations, e.g. [77], showed that in the context of wind power predictions the use of the predicted wind vector from the diagnostic level at 10 m height leads to better results than input from the genuine model levels at 33 m or 110 m. This is surprising at first glance because the wind field at a diagnostic level is derived from higher levels by a parametrisation of the wind profile rather than a solution of the equations of motion. However, this parametrisation includes the logarithmic wind profile and stability corrections and seems to lead to a more suitable input for the procedures that are used afterwards to transform wind speed to the power output of wind farms.

The DM predictions used here are from the years 1996, 1997 and 1999, where the focus is on 1996 due to the highest data availability in this year. The data of the year 1999 differ from those of the other years as the predictions are provided as point predictions that are already interpolated to the location of the wind farm by the weather service. Moreover, the wind speed prediction is given in a resolution of 1 knot (approximately 0.5 m/s) compared with 0.1 m/s in 1996–1997.

5.3 Measurements

5.3.1 Wind Data and Power Output of Wind Farms

In the framework of the WMEP programme (Scientific Measurement and Evaluation Programme) funded by the German federal government, the electrical power output of wind turbines has been recorded on a regular basis since 1990. In addition, wind speed and wind direction are measured at either 10 m or 30 m height on a nearby meteorological mast. The time series are sampled in 5-min intervals, which is more than sufficient for the investigations in this work. As discussed above the time series of all involved quantities are averaged over 1 h to make them comparable to those of the NWP. This means that the power output is effectively integrated over 1 h, so that it corresponds to the amount of electrical energy produced in this time period. Moreover, the averaged values of wind speed and direction are also used to assess the local meteorological conditions (Chap. 9).

In the remaining chapters, measurements from about 30 WMEP sites are used. The map in Fig. 5.1 shows the locations of the sites, mainly from the northern half of Germany, considered in this work. As availability and quality of the data varies significantly from site to site and among different years, most of the following investigations will be restricted to a subset of the total set of stations.

5.3.2 Meteorological Mast in Cabouw

For detailed investigations of the applicability of the Monin–Obukhov theory for wind power predictions in Chap. 7, measurements from a met-mast in Cabouw, the Netherlands, are used. The measurements are provided by the Royal Netherlands Meteorological Institute (KNMI). The following data have been used:

- wind speed and direction at 10, 20, 40, 80, 140 and 200 m height
- temperature at 2, 10, 20, 40, 80, 140 and 200 m height

The measurement data are available from January 1993 to December 1996 with a temporal resolution of 30 min and from May 2000 to March 2002 as 5-min values. The concrete mast has a diameter of 2 m and a total height of 213 m. At each measurement height, three sensor-carrying booms with a length of 10.5 m are mounted pointing in different directions. Thus, according to wind direction and availability, different time series can be used and shadowing effects of the mast can be reduced. A detailed description can be found in [6].

To ensure a homogenuous roughness, only the wind directions from the 180° sector (165–195°) are considered. Wessels [120] recommended this sector because the flow distortion of the mast is rather low for these wind directions. The roughness length in this sector is set to 0.01 m. Situations with very strong wind speed gradients (>0.7 m/s within 30 min) or temperature gradients (>0.5 K within 30 min) are neglected. In addition, situations with wind speeds lower than 2.5 m/s, which are not relevant for wind energy applications, are not considered.

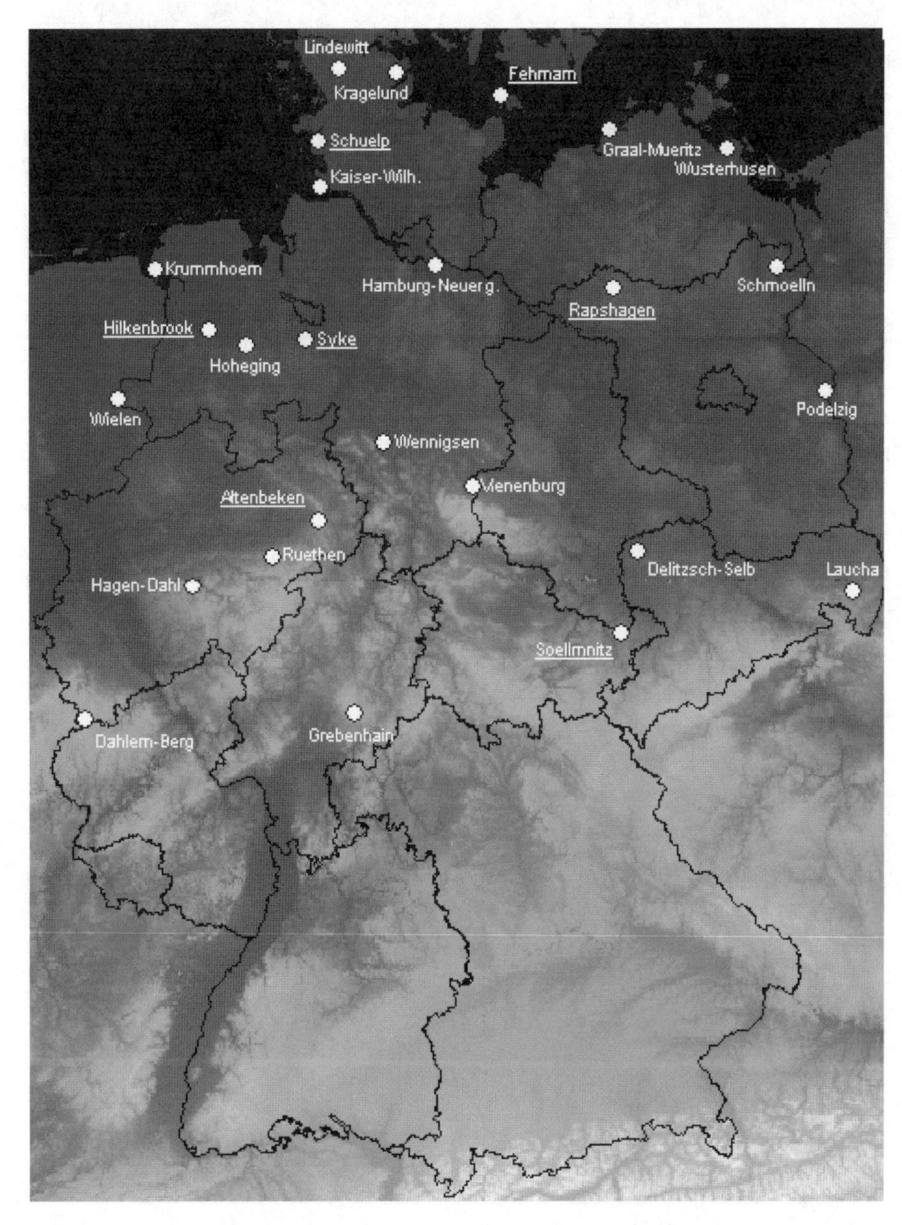

Fig. 5.1. Set of 30 WMEP sites in Germany with measurements of wind speed, wind direction and power output of wind turbines that are used to verify the predictions and to assess the local meteorological conditions. The underlined stations are used for detailed investigations in Chaps. 6 and 9

5.3.3 Atmospheric Pressure

Measurements of atmospheric pressure will be considered in the assessment of the meteorological conditions at a site. However, as the WMEP sites are not equipped with readings of surface pressure, the measurements from the nearest synoptic station of the German Weather Service (DWD) are used. The distance between the sites and the synoptic stations are in the range 5–30 km, but this is regarded as uncritical because horizontal gradients of the pressure vary only little on this scale. In contrast to the WMEP data the pressure time series are recorded on an hourly basis which is appropriate to account for changes on a synoptic scale.

To normalise all sites to a common pressure level the surface pressure, $\mathrm{p_{surface}}$, at ground level is corrected to the pressure at mean sea level (pmsl) using the barometric height formula

$$\mathrm{pmsl} = \mathrm{p_{surface}}\, e^{\frac{gh}{RT}} , \tag{5.1}$$

where g is the gravitational constant, h is the height of the synoptic station above mean sea level, R is the dry gas constant and T is the surface temperature which is also measured at the synoptic station.

6

Assessment of the Prediction Accuracy

Abstract. This chapter gives an overview of different aspects of the prediction accuracy. The accuracy of the predictions is assessed by comparing predicted wind speeds and power outputs with corresponding measurements from a selection of six out of 30 sites in Germany. Starting from a visual inspection of the time series the statistical behaviour of the forecast error is investigated showing strong evidence that the differences between predicted and measured wind speed are normally distributed at most sites, while the distribution of the power differences is far from Gaussian. To quantitatively assess the average forecast error the root mean square error (rmse) is decomposed into different parts which allow to distinguish amplitude errors from phase errors. The analysis shows that amplitude errors can mainly be attributed to local properties at individual sites, while phase errors affect all sites in a similar way. The relative rmse of the power prediction is typically larger than that of the wind speed prediction by a factor of 2 to 2.5. It turns out that this is mainly caused by the increased relative amplitude variations of the power time series compared to the wind time series due to the nonlinear power curve. In addition, the cross-correlation is virtually not affected by transforming wind speed predictions to power output predictions. Hence, phase errors of the wind speed prediction are directly transfered to the power prediction. Moreover, the investigation indicates that there is little space for correction schemes that are based on linear transformations of the complete time series to substantially improve the prediction accuracy.

6.1 Introduction

Predictions of the future development of meteorological variables are not perfect, and this is continuously confirmed by everyday experience and scientific investigations. Hence, in order to use and to improve forecasting systems the quality of the predictions has to be evaluated, where "quality" refers to a judgement of how good or bad the prediction is. For this purpose the predicted values are typically compared with the corresponding measurements. In the case of continuous variables such as wind speed the easiest way to get an idea of the quality of the forecast is by plotting the two time series and visually assessing the deviations between them, which is used in this chapter to illustrate typical errors that can occur.

In general, the quantitative assessment of the relationship between forecast and prediction involves the use of standard statistical methods that will be referred to as “error measures”. These error measures are based on calculating a suitable average over the deviations between predicted and measured values over a certain time period either by using the straightforward difference between the two or by taking the squared difference to eliminate the signs. Hence, in this work the term “error” refers to the numerical value found by applying one of the error measures to the predicted and measured time series. However, note that the difference between prediction and measurement is denoted as pointwise error.

Of course, using the error measures requires that the data have already been recorded, i.e. the error is always a historical value representing the forecast quality of the past. The uncertainty, on the other hand, is understood as the expected error of future predictions which is unknown a priori. Under the assumption that the statistics of the errors are stationary, the historical error is used as an estimate of uncertainty.

In order to interpret error and uncertainty as confidence intervals, i.e. as a certain range around the predicted value in which the measured value lies with a well-defined probability, the underlying distribution of the differences between prediction and measurement has to be known. Therefore, the statistical distributions of the pointwise prediction errors are investigated for wind speed and power predictions.

6.2 Basic Visual Assessment

A graphical representation of the predicted and measured data conveys a first impression of the forecast accuracy. In Figs. 6.1 and 6.2 the time series of the power prediction of a single wind turbine in the North German coastal region over an interval of 6 days are compared with the corresponding WMEP measurement data with an hourly resolution. The overall agreement between the two time series is rather good in the period of time shown in Fig. 6.1, while the sample in Fig. 6.2 illustrates a poor forecast accuracy.

This example highlights two characteristic sources of error occurring in the forecasting business: deviations in amplitude, i.e. overestimation or underestimation by the forecast but with a correct temporal evolution (as on day 254 in Fig. 6.2), and phase errors, i.e. the forecast would match the real situation if it were not shifted in time (day 257). The following statistical investigation has to account for these effects and must, therefore, be based on error measures that quantitatively assess the amplitude and phase errors.

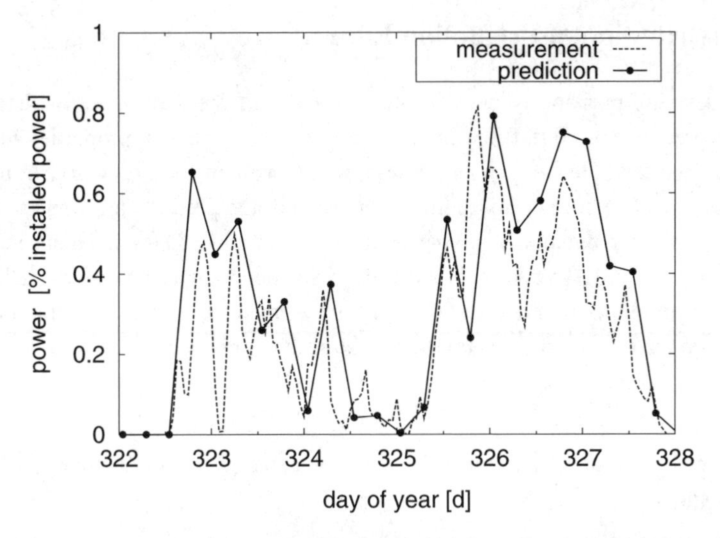

Fig. 6.1. Comparison of the time series of measured and predicted (6–24 h) power output normalised to rated power at one site. The agreement between the two time series is rather good over the period shown. In particular, the increase in wind speeds on days 323 and 325 is correctly predicted. However, the amplitudes, especially on day 326, do not completely fit

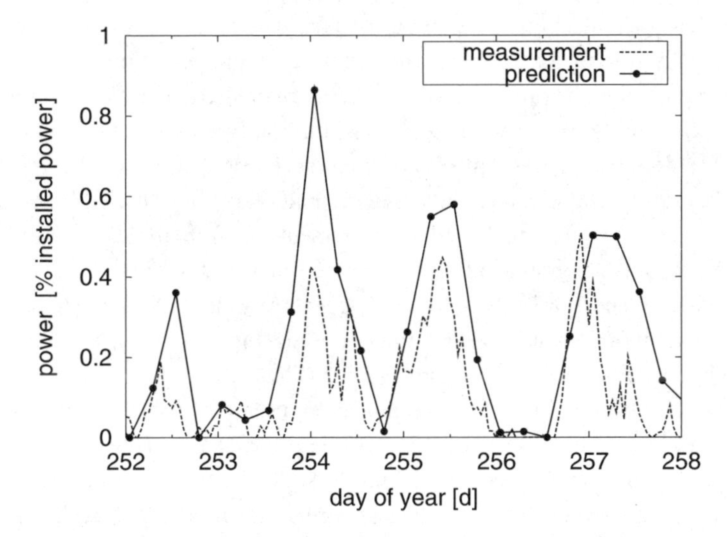

Fig. 6.2. Comparison of the time series of measured and predicted power output normalised to rated power at one site but for a period of time with a rather poor agreement. Two characteristic errors can be observed. On day 254 a typical amplitude error occurs, where the prediction is in phase with the measurement but strongly overestimates the real situation. In contrast to this, on day 257 the amplitude of the prediction is right but the maximum wind speed appears several hours earlier and decays faster than predicted. Hence, this situation is an example for a phase error

6.3 Distribution of Prediction Errors

The underlying probability density function (pdf) of the forecast error determines the interpretation of confidence intervals and further statistical properties in the remaining chapters. Hence, prior to assessing the error of the prediction in terms of statistical measures, it is important to analyse how the prediction errors are distributed. In particular, the question whether the error follows a Gaussian distribution has to be tested carefully. In this section the error is understood as the difference between prediction and measurement. Let $x_{\mathrm{pred},i}$ be the predicted and $x_{\mathrm{meas},i}$ the measured value; then the deviation between the two at time i is given by

$$\epsilon_i := x_{\mathrm{pred},i} - x_{\mathrm{meas},i} \ . \tag{6.1}$$

This is the definition of the pointwise error, which is the basic element in the error assessment.

6.3.1 Wind Speed Prediction

The wind speed supplied by the numerical weather prediction (NWP) model of the German Weather Service (DWD) is the main input into the power prediction system and has a major impact on the accuracy. As forecast errors are expected to change systematically with increasing forecast horizon, each prediction time t_{pred} is treated separately. Dividing the differences, $\{\epsilon_i\}$, between prediction and measurement into bins and counting the relative frequency within the bins leads to the an empirical probability density function pdf(ϵ). For wind speed predictions of the years 1996, 1997 and 1999, pdf(ϵ) is calculated based on predictions from the DM model of the DWD and the corresponding WMEP measurement data from the same period of time (Chap. 5). Results are graphically shown for selected sites in Fig. 6.3.

The visual inspection of these figures suggests that most sites seem to have a normal distribution of the wind speed prediction error (Fig. 6.3 (top)), while the type of distribution for other sites is not clear (Fig. 6.3 (bottom)). A close to Gaussian distribution has also been inferred from graphical representations of the wind speed prediction error by Giebel [35] and Landberg [58] for the Danish NWP model *HIRLAM*.

As pointed out earlier it is very helpful to know the type of error distribution to interpret the standard deviations of these distributions in terms of confidence intervals. This calls for a more detailed analysis of the characteristics of the many sites and prediction times. Consequently, the error distributions are checked for normality by using standard statistical tests. All distributions are run through a parametrical χ^2 test and a non-parametrical Lilliefors test with the hypothesis "pdf(ϵ) is normally distributed" at a typical significance level of 0.01. This means that with a probability of 1% the hypothesis is falsely rejected although it is correct. Details on both testing methods are given in Appendix B.

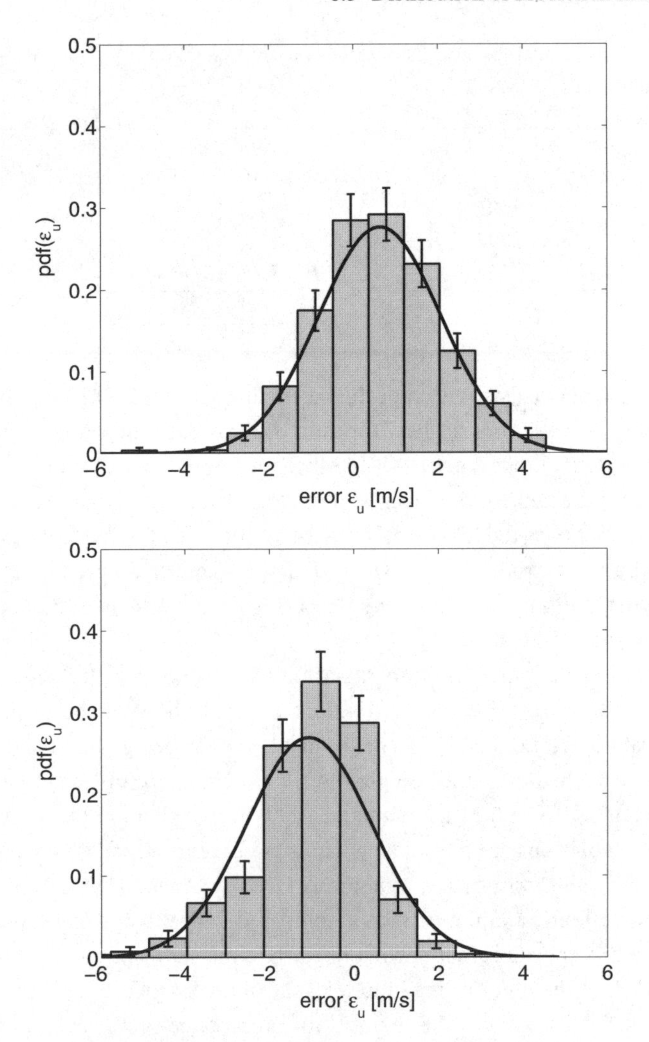

Fig. 6.3. Probability density of the deviations between predicted and measured wind speeds at 10 m height with 12 h lead time in the year 1996. A normal distribution with the same mean and standard deviation is given by the solid line. The error bars illustrate the 68% confidence levels. For the site Fehmarn (*top*) the distribution seems to be close to Gaussian, while the errors of prediction in Altenbeken (*bottom*) deviate from normality

The test results for the years 1996 and 1997 show that a majority of the predictions have normally distributed errors. A total of 120 tests, i.e. 20 sites with 6 prediction times each, are performed for each year. A summary is given in Table 6.1. In 1996 about 82% of these tests do not reject the hypothesis of a normal distribution; in 1997 the rate is even higher at 93%. The consistency of the results between the two testing methods is rather good. In 1996 nine out of 20 tested stations pass

Table 6.1. Results of testing error distributions of wind speed for normality using the χ^2 test and the Lilliefors test

Year	Number of tests	Not rejected		
		χ^2 test (%)	Lilliefors test (%)	simult. all prediction times (%)
1996	120	82	81	45
1997	120	92	93	65
1999	256	69	63	13
Total	496	77	74	43

both tests simultaneously for each of the six prediction times. The same holds for 14 out of 20 sites in 1997. For all lead times in the two years in a row seven sites still have close to normal distributions. Nevertheless, the number of failures in the tests is higher than expected for the given significance level. A closer look at the distributions that do not pass the test reveals that their pdfs are systematically different (as in Fig. 6.3 (bottom)) from those of a normal distribution for all prediction times. The reason for this, although, not clear, might be systematic effects in the measurement procedure or local flow distortion.

In contrast to the previous years the 1999 data do not pass the tests that easily: 69% of the 256 χ^2 tested error distributions and only 61% of the Lilliefors tested distributions were not rejected. There are just two sites being simultaneously tested positive by both methods at all prediction times. The main difference between the years 1996 and 1997 on the one hand and 1999 on the other are the prediction data. In 1999 the predictions are provided as point predictions already interpolated to the location of the wind farm by the weather service. Moreover, the wind speed prediction is given with a resolution of 1 knot compared with 0.1 m/s in 1996–1997. This has an impact on the statistical behaviour of the time series because 1 knot is about 0.5 m/s, which of the order of magnitude of the observed effect.

Hence, for 1996 and 1997 a majority of pdfs of the wind speed prediction error can be reasonably well described by a normal distribution. Primarily, this is a convenient property as the standard deviations can be interpreted as 68% confidence intervals. Moreover, as the auto-correlation function of the time series of deviations between prediction and measurement of a specific lead time decays very rapidly, the prediction errors at the same prediction times on succeeding days can be regarded as statistically independent. This means that in the following statistical investigations the time series data can be considered as a set of independent samples which, e.g., simplifies the calculation of errorbars.

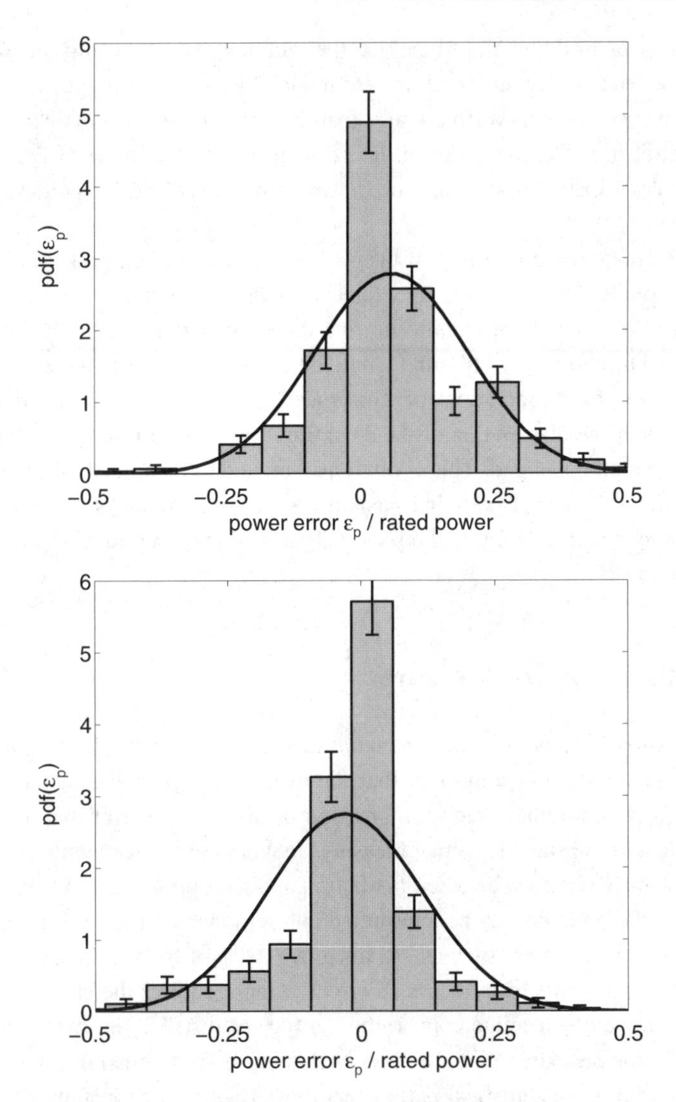

Fig. 6.4. Probability density function of the forecast error of the power prediction in the year 1996 for the same sites and prediction times as in Fig. 6.3. The error ϵ_p is normalised to the rated power of the wind turbine. For both sites the distributions are far from being normal

6.3.2 Error in Power Prediction

In contrast to errors in the wind speed prediction, the statistical distributions of the power prediction error are completely different. Figure 6.4 indicates that the error pdfs are unsymmetric and non-Gaussian with, in particular, higher contributions to small values, especially near zero. This is related to the fact that wind speeds below the cut-in speed of the wind turbine, i.e. the minimum wind speed (typically around 4 m/s) that leads to a power output, are mapped to zero by the power curve such that this event occurs more frequently than before. Moreover, the distributions are unsymmetric in most cases. Hence, it is not surprising that none of the 496 tested distributions at different sites and prediction times passes the hypothesis of being normal.

To determine the proportion of events inside the confidence interval, the pdf of the power prediction error was integrated over the σ interval around the bias. For the majority of sites the probability to find the error in this interval is 77%. This is profoundly larger than the 68% that would be expected if the pdf were normal.

So by converting predictions of wind speed to wind power the statistical properties in terms of the distribution of the deviations between forecast and measurement are fundamentally changed. This is obviously related to the impact of the non-linear power curve as the key element in translating wind speed to power output. In Chap. 8 the mechanism that transforms the pdfs is described and used to model the effect of the power curve.

6.4 Statistical Error Measures

Each meteorological prediction system has to prove its quality according to standard statistical methods. It is important that the evaluation procedure of the prediction accuracy provide a rather precise impression of the average error that has to be expected. Most of the statistical error measures inevitably produce numbers that somehow assess the deviations between prediction and measurement. But it is, of course, crucial that these results can be interpreted in a reasonable manner. It is desirable to come up with a statement such as "At a site XY 68% of the wind speed predictions deviate less than 1 m/s from the mean error", which requires the choice of the right error measure and some clue concerning the type of distribution. In this section the statistical error measures that are used in this work are defined and motivated focusing only on those statistical parameters that provide some insight into the error characteristics rather than trying out all measures that are available. As the assessment of the error plays an important role throughout the remainder of this work, the statistical error measures are discussed here in greater detail instead of being exiled into the appendix.

A further useful quality check is, of course, whether the prediction system performs better than any trivial type of forecasting technique such as persistence, or the climatological mean or forecasts provided by *The Old Farmer's Almanac* [110]. To put it differently, the prediction system has to be evaluated against a simple reference system. For this purpose various skill scores have been developed [116]. In this work the predictions are compared to persistence as a reference system, but this is not pursued in detail.

In what follows, statistical error measures describe the average deviations between predicted and measured values. The average is normally taken over one year to include all seasons with a chance of covering most of the typical meteorological situations. The error measures commonly used to assess the degree of similarity between two time series are based on the difference between prediction and measurement according to (6.1), i.e. $\epsilon_i := x_{\mathrm{pred},i} - x_{\mathrm{meas},i}$.

6.4.1 Decomposition of the Root Mean Square Error

The root mean square error (rmse) is rather popular among the many measures that exist to quantify the accuracy of a prediction. Despite being considered as a rather rough instrument, it is shown in this section how the rmse can be split into meaningful parts which shed some light on the different error sources. With the definition in (6.1) the rmse between the two time series x_{pred} and x_{meas} is defined by

$$\mathrm{rmse} := \sqrt{\overline{\epsilon^2}} \,, \tag{6.2}$$

where the overbar denotes the temporal mean.

The rmse can easily be expressed in terms of the *bias* and the *variance of the error* (see Appendix A for detailed definitions). Using simple algebraic manipulations the error variance can further be separated into two parts, where one is more related to amplitude errors and the other more to phase errors. This decomposition has been beneficially used in previous investigations, e.g. by Hou et al. [46] or Takacs [108]. Hence, with the notation from Hou et al. [46], the decomposition of the rmse is given by

$$\begin{aligned} \mathrm{rmse}^2 &= \mathrm{bias}^2 + \mathrm{sde}^2 \\ &= \mathrm{bias}^2 + \mathrm{sdbias}^2 + \mathrm{disp}^2 \,, \end{aligned} \tag{6.3}$$

where

$$\mathrm{bias} = \overline{\epsilon} \,,$$

$$\mathrm{sde} = \sigma(\epsilon) \,, \tag{6.4}$$

$$\mathrm{sdbias} = \sigma(x_{\mathrm{pred}}) - \sigma(x_{\mathrm{meas}}) \,, \tag{6.5}$$

$$\mathrm{disp} = \sqrt{2\sigma(x_{\mathrm{pred}})\sigma(x_{\mathrm{meas}})(1 - r(x_{\mathrm{pred}}, x_{\mathrm{meas}}))} \,, \tag{6.6}$$

with $r(x_{\mathrm{pred}}, x_{\mathrm{meas}})$ denoting the cross-correlation coefficient between the two time series and $\sigma(x_{\mathrm{pred}})$ or $\sigma(x_{\mathrm{meas}})$, respectively, denoting their standard deviations. Detailed definitions of the statistical quantities are given in Appendix A.

Equation (6.3) connects the important statistical quantities of the two time series. It shows that three different terms contribute to the rmse originating from different effects. The bias accounts for the difference between the mean values of prediction and measurement. The standard deviation, sde, measures the fluctuations of the error around its mean. As seen in Sect. 6.3, sde is very useful as it directly provides 68% confidence interval if the errors are normally distributed. In the context of comparing prediction and measurement, sde has two contributions, First, sdbias, i.e. the difference between the standard deviations of x_{pred} and x_{meas}, evaluates errors due to wrongly predicted variability. This is together with the bias, an indicator for amplitude errors. Second, the dispersion, disp, involves the cross-correlation coefficient weighted with the standard deviations of both time series (6.6). Thus, disp accounts for the contribution of phase errors to the rmse.

Takacs [108] used the decomposition (6.3) in the context of numerical simulations of the advection equation with a finite-difference scheme. The rmse between the numerical solution on a grid and the true analytical solution was split into two parts. On the one hand bias^2+sdbias^2 is related to numerical dissipation, which means that energy is lost due to the finite-difference formulation of the equation of motion. In the context of this work, dissipation refers to the more general phenomenon that the amplitudes of the predicted and observed time series are systematically different. On the other hand Takacs used disp^2, which "increases due to the poor phase properties" [108].

6.4.2 Limits of Linear Correction Schemes

The prediction accuracy can in many cases be substantially improved by eliminating systematic errors as much as possible. It is, of course, desirable, to know the reasons for these errors and correct them using physical modelling, but often the cause for systematic errors cannot be pinned down exactly. This is where statistical methods which describe the overall characteristics of the errors come in and allow for global corrections of the time series. Such a post-processing is often referred to as model output statistics (MOS). A straightforward approach in this direction is a linear transformation of the predicted values such that they on average match amplitude and offset of the measured time series better than before.

However, any linear correction applied to the time series leaves the cross-correlation coefficient $r(x_{\mathrm{pred}}, x_{\mathrm{meas}})$ unaffected. Therefore, the cross-correlation is regarded as the "king of all scores" [49] in weather and climate forecasting. Due to this invariance the dispersion, disp, cannot easily be reduced by linear manipulations of the time series, this limits, of course, the space for improvements in the prediction accuracy with statistical corrections such as MOS that are based on linear transformations.

The bias and sdbias are sensitive to linear manipulations of the time series. Hence, if an improvement of the forecast method leads to a better performance in terms of these error measures, it can be concluded that systematic errors mainly related to the amplitude of the prediction have been removed. In contrast to this, changes to the prediction method that positively affect disp can be seen as substantial improvement in the forecast quality of the temporal evolution.

In the following, two slightly different ways of finding a correction are described that provide some benefit in terms of the overall accuracy and the understanding of the error. Both approaches are based on a linear transformation of the original prediction to obtain an improved forecast. Hence,

$$\tilde{x}_{\text{pred}} := \alpha x_{\text{pred}} + \beta \,, \tag{6.7}$$

where x_{pred} is the prediction, while α and β are real numbers. The influence of this transformation on the statistical error measures introduced in the previous section will be discussed and the maximum decrease in rmse that can be obtained will be calculated.

Linear Regression

A popular and successfully applied type of post-processing is based on linear regression of the data. This method aims at minimising the rmse between the linearly transformed prediction (6.7) and the measurement, i.e.

$$\sqrt{\sum(\alpha x_{\text{pred}} + \beta - x_{\text{meas}})^2} \rightarrow \min \,. \tag{6.8}$$

This condition gives non-ambiguous solutions for α and β [12]:

$$\begin{aligned} \alpha &= \frac{\sigma(x_{\text{meas}})}{\sigma(x_{\text{pred}})} \mathrm{r}(x_{\text{pred}}, x_{\text{meas}}) \,, \\ \beta &= \overline{x_{\text{meas}}} - \alpha \overline{x_{\text{pred}}} \,. \end{aligned} \tag{6.9}$$

All quantities on the right-hand side of (6.9) that are needed to calculate the parameters can be estimated from the two time series x_{pred} and x_{meas}. Naturally, a certain number of data points has to be considered to ensure statistically significant parameters.

The implications of this transformation for the statistics of the corrected prediction are easy to see by calculating the standard statistical measures:

$$\begin{aligned} \text{bias}_{\text{LR}} &= 0 \,, \\ \text{rmse}_{\text{LR}} &= \sigma(x_{\text{meas}}) \sqrt{(1 - \mathrm{r}^2(x_{\text{pred}}, x_{\text{meas}}))} \,. \end{aligned} \tag{6.10}$$

As linear regression by definition minimises the rmse, the expression in (6.10) is the lower boundary of the error that can be achieved by linear transformation of the time series.

If the transformation (6.7) is applied to the same set of data points that were used to estimate α and β, the maximum reduction of the rmse is obtained. Of course, the idea is to use the parameters of historical data to correct future predictions. Then the improvement of the forecast error strongly depends on the stationarity of α and β.

Double Bias Correction (DBC)

The decomposition of the rmse in (6.3) suggests a linear correction of the prediction that simultaneously eliminates bias and stdbias from the rmse. In this work this transformation will be denoted as *double bias correction* (DBC). The conditions

$$\begin{aligned} \mathrm{bias}_{\mathrm{DBC}} &= 0 \,, \\ \mathrm{sdbias}_{\mathrm{DBC}} &= 0 \end{aligned} \tag{6.11}$$

lead to the transformation parameters

$$\begin{aligned} \alpha &= \sigma(x_{\mathrm{meas}})/\sigma(x_{\mathrm{pred}}) \,, \\ \beta &= \overline{x_{\mathrm{meas}}} - \alpha\overline{x_{\mathrm{pred}}} \,, \end{aligned} \tag{6.12}$$

which, in contrast to the linear regression parameters in (6.9), do not involve the correlation coefficient between the two data sets.

With the conditions in (6.11) the rmse of the corrected prediction is then obviously identical to the dispersion

$$\begin{aligned} \mathrm{rmse}_{\mathrm{DBC}} &= \mathrm{disp}_{\mathrm{DBC}} \\ &= \sigma(x_{\mathrm{meas}})\sqrt{2(1 - \mathrm{r}(x_{\mathrm{pred}}, x_{\mathrm{meas}}))} \,. \end{aligned} \tag{6.13}$$

Comparison of Both Methods

The linear regression method provides the lowest possible forecast error that is achievable with linear manipulations of the global time series as it is based on linear regression, which by construction minimises the rmse.

Linear regression uses the variability and the cross-correlation of the two time series to rescale the prediction, i.e. this method exploits their statistical dependence. In contrast to this, DBC only refers to the variabilities which are statistically independent. In fact it rescales the prediction to the statistical properties of the observations. Thus, apart from a factor $\sigma(x_{\mathrm{meas}})$, DBC is equivalent to normalising both time series separately with their respective means and standard deviations. Note that both

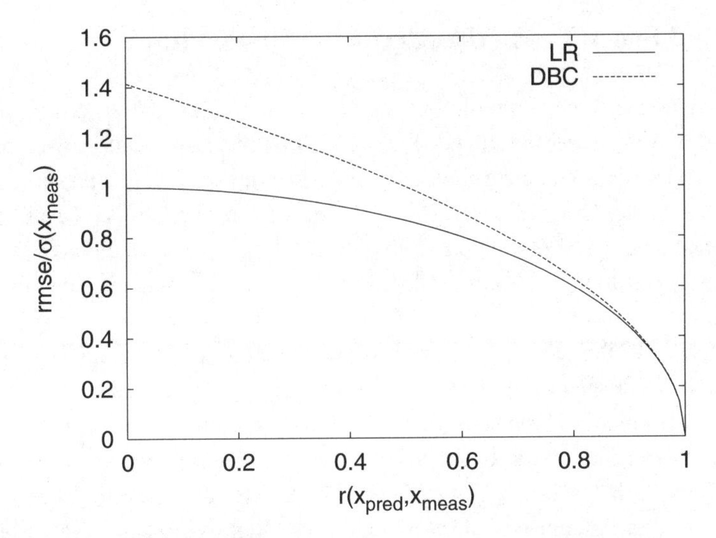

Fig. 6.5. Root mean square error (rmse) normalised by $\sigma(x_{\mathrm{meas}})$ for the correction based on linear regression (LR) and the double bias correction (DBC) versus cross-correlation coefficient, $r(x_{\mathrm{pred}}, x_{\mathrm{meas}})$, between prediction and measurement. The difference between the two methods is small for r close to 1

types of correction leave the cross-correlation unchanged as it is invariant under linear transformations.

The rmse normalised by $\sigma(x_{\mathrm{meas}})$ for the two corrections is shown in Fig. 6.5. It only depends on the cross-correlation. The difference between both methods decreases with increasing correlation. For $r > 0.8$, $\mathrm{rmse}_{\mathrm{DBC}}$ and, thus, $\mathrm{disp}_{\mathrm{DBC}}$ are less than 5% larger than $\mathrm{rmse}_{\mathrm{LR}}$. This means that the dispersion already provides a good estimate of the lower boundary of the rmse in terms of linear corrections.

The results provided here are derived for an a posteriori correction of data, i.e. the data of one year are corrected with the parameters of the same year. The idea of any correction is, of course, to apply parameters that were estimated from historical data to future predictions. This inherently assumes that these statistical parameters are quasi stationary and do not change much in time. In practice this condition is not perfectly fulfilled, so that the improvements of the rmse given in (6.10) and (6.13), respectively, are lower bounds to what can be expected for future predictions.

As a consequence of these considerations the verification of the predictions given in this work will involve the rmse divided into the three parts given in (6.3). The prediction accuracy of wind speed and power output will be analysed separately focusing on their typical statistical behaviour at different locations.

6.5 Wind Speed Prediction Error for Single Sites

For the assessment of the prediction accuracy at individual sites the time series of predicted and measured wind speeds as described in Chap. 5 taken over one year are used to calculate the error measures discussed above. The results shown in this section are mainly from the year 1996, as in this year data quality and availability of both measurement and prediction are higher for a majority of stations than in 1997 and 1999. Nevertheless, the effects described here have been investigated using data from all three years.

From the 30 sites available in 1996, six with a rather high data quality and availability have been selected as test cases to illustrate the typical statistical behaviour at different locations. Three sites are situated in flat terrain in the north of Germany: Fehmarn on an island in the Baltic Sea, Schuelp at the North Sea Coast and Hilkenbrook about 70 km away from the coast. Three more sites can be found in more complex terrain: Altenbeken and Soellmnitz in a slightly mountainous terrain, and Rapshagen with less but still noticeable orography. These test cases are underlined in Fig. 5.1, where the site Syke will be used in a later investigation.

The following analysis focusses on the general properties of the prediction error of the wind speed at the selected stations. The aim is to identify statistical characteristics that are global, i.e. similar for different sites, and local, i.e. originating from on-site conditions. Therefore, the error measures will be normalised to the mean measured wind speed. In Table 6.2 these mean values are given together with the standard deviation of the measured wind speed. As the ratio between both the quantities is in a rather narrow range between 0.50 and 0.60, the standard deviation is roughly proportional to the annual average of the wind speed.

The standard deviation of the wind speed will be considered in the following as it plays an important role with regard to the error measures, in particular in the sdbias

Table 6.2. Annual mean values of measured wind speeds for the selected sites and corresponding standard deviations of the wind speed[a]

Site name	Mean measured wind speed, $\overline{u_{\mathrm{meas}}}$ (m/s)	Standard deviation of measured wind speed, $\sigma(u_{\mathrm{meas}})$ (m/s)	Relative standard deviation $\sigma(u_{\mathrm{meas}})/\overline{u_{\mathrm{meas}}}$
Altenbeken	4.4	2.5	0.57
Fehmarn	5.6	3.0	0.54
Hilkenbrook	3.5	2.1	0.60
Rapshagen	4.2	2.1	0.50
Schuelp	4.9	2.6	0.53
Soelmnitz	3.6	1.9	0.53

[a] The ratio between the two is between 0.50 and 0.60, i.e. mean wind speed and standard deviation are roughly proportional.

(6.5) and disp (6.6). However, note that the wind speed follows a Weibull distribution and, hence, the standard deviations of the wind speed cannot be interpreted as 68% confidence intervals. They are used here as a measure of the fluctuations of the time series.

The rmse of the wind speed prediction and the different contributions to it according to the decomposition in (6.3) are shown in Fig. 6.6–6.10. All values are normalised to the mean measured wind speed given in Table 6.2. In the lower left corner of each plot the estimated error bar related to the calculation of the corresponding statistical quantity (rmse, bias, sde, sdbias, dispersion and correlation) based on a finite time series is shown. To avoid overloaded plots they were not drawn around each data point. The error bars are estimated by randomly dividing the time series in M equally sized subsets. The statistical error measures are then calculated for each subset separately and the standard deviation of the M results is determined. For different M this standard deviation is divided by $\sqrt{M}$ assuming that the M subsets are uncorrelated. This allows for an extrapolation to $M = 1$, which is an estimation of the error bar for the complete time series. In most cases the estimated values are rather consistent. For this investigation the error bars were chosen to represent the largest estimation found with this procedure.

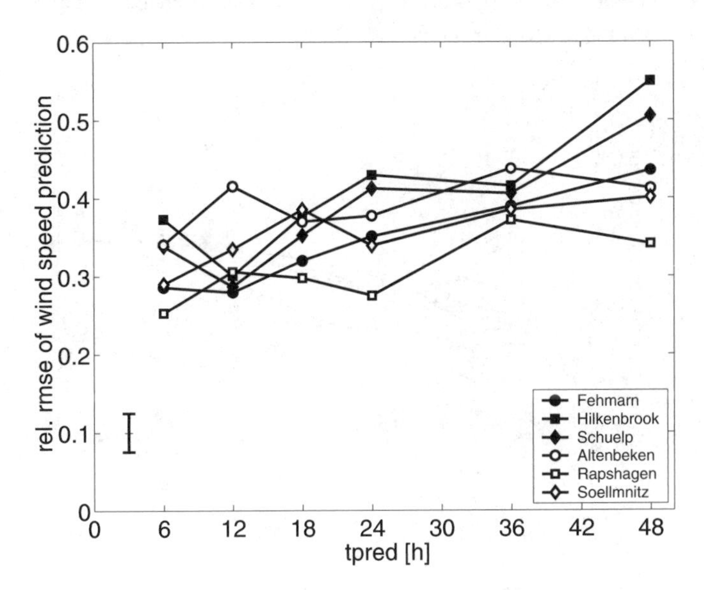

Fig. 6.6. Root mean square error (rmse) between predicted and measured wind speed at 10 m height for the selected sites in 1996 normalised to annual mean wind speed according to Table 6.2. The rmse is calculated separately for the different prediction times and seems to increase with increasing forecast horizon. There is no apparent coherent behaviour among the different sites. The error bar in the lower left corner is valid for all data points

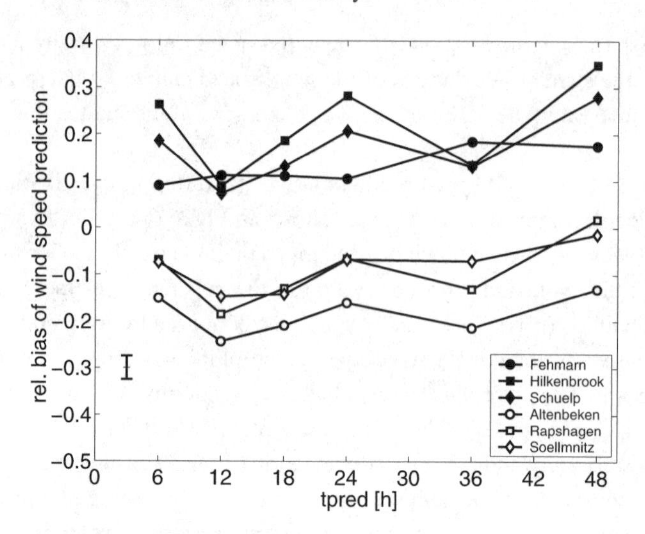

Fig. 6.7. Relative bias of the mean values of predicted and measured wind speed for the same sites as in Fig. 6.6. Most sites show "W"-shaped diurnal variations with a comparable amplitude. One exception is the island site Fehmarn with only weak differences between daytime and nighttime bias. Sites located in flat terrain (*solid symbols*) have positive bias (Fehmarn, Schuelp, Hilkenbrook), while sites in complex terrain (*open symbols*) tend to have negative bias (Altenbeken, Rapshagen, Soellmnitz)

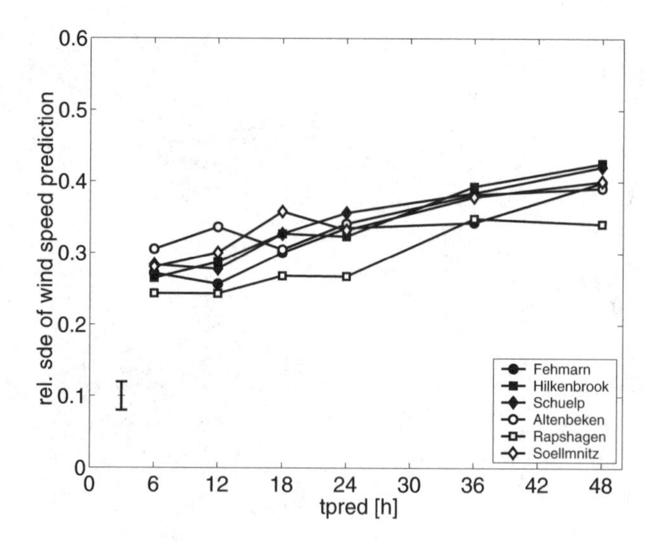

Fig. 6.8. Relative standard deviation of the error, sde, for the selected sites. Increase of sde with prediction time with diurnal effects is still noticeable. These values directly provide the 68% confidence interval if the underlying error distributions are normal

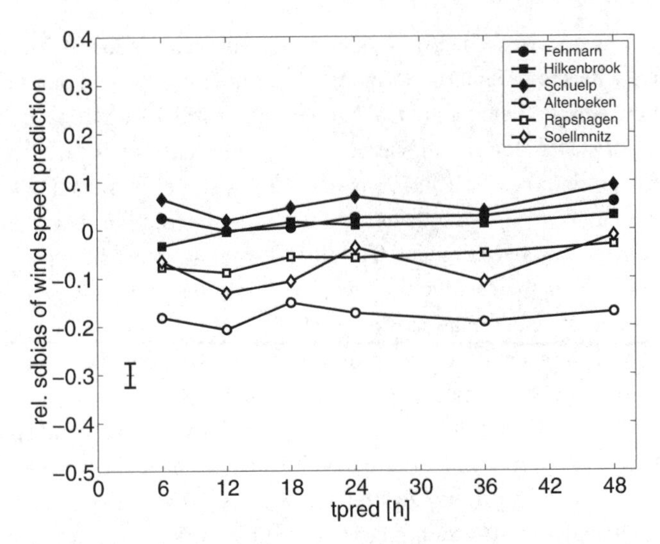

Fig. 6.9. Difference between standard deviations of predicted and measured time series (sdbias) normalised with the mean wind speed for the selected sites. As in Fig. 6.7, the three flat terrain sites (*solid symbols*) group together in the range of small sdbias unlike the behaviour of Altenbeken with a strongly negative sdbias

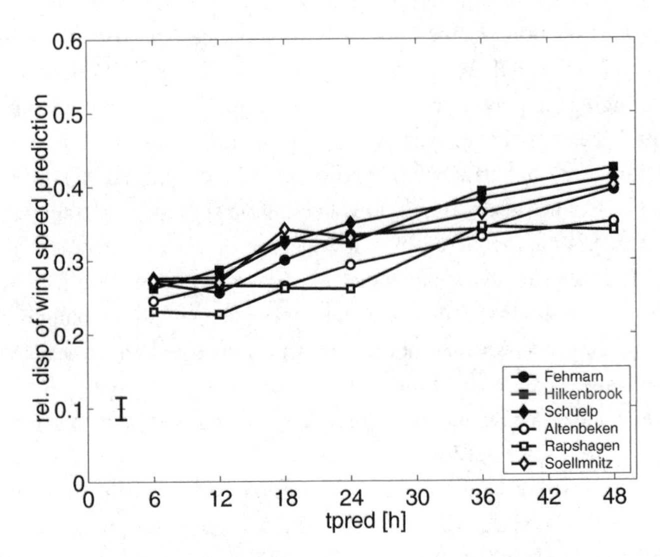

Fig. 6.10. Relative dispersion of the wind speed error versus prediction horizon. The relative dispersion is in a rather narrow range for the selected sites. It increases with prediction time with a similar rate at all sites indicating a systematic error that affects different sites in the same way

Starting with the rmse in Fig. 6.6, one can observe a general increase of the relative rmse with the prediction horizon t_{pred}, which is expected due to a growing mismatch between predicted and "real" situations. Diurnal variations are noticeable but they are not systematic for all sites; e.g. some sites have relatively larger rmse at noon, i.e. prediction times 12 h and 36 h, than at other times, while it is vice versa for others. The relative rmse of the different sites are rather similar, being in the range 0.25–0.37 for the 6-h prediction and 0.34–0.55 for the 48-h prediction.

The behaviour of the relative bias, i.e. the difference between the mean values of prediction and measurement, as shown in Fig. 6.7, is consistent for the majority of sites. Most of the sites have strong diurnal variations indicating that on average the relative bias at noon is smaller than at other times. The differences between night and day (12 h and 24 h) are of the order of 10–20% of the mean wind speed, which corresponds to 0.4–0.7 m/s. All investigated inland sites have this typical "W" shape of the bias, while for the island site Fehmarn the diurnal variations are relatively weak. Sites located in flat terrain have bias with positive values for all prediction times, while all sites that can be attributed to complex terrain come up with negative bias at almost all t_{pred}. Therefore, in contrast to the relative rmse, sites in different terrain type are clearly separated with regard to the relative bias. This is due to the fact that the sign of the bias, which is neglected in the rmse (see (6.3)), is important.

The diurnal variations in the bias have already been described by Focken [27] and Mönnich [77] and will be further investigated in Chap. 7. Differences between daytime and nighttime bias clearly suggest a systematic error in describing stability effects of the lower boundary layer by the NWP model. This phenomenon is related to the coupling between surface wind and wind in higher altitudes. During the night the land surface cools the air from below, leading on average to a stable atmospheric stratification that is characterised by a low degree of vertical turbulent transport of momentum. In this case the surface wind is only weakly coupled to its driving force, which is the wind in higher altitudes. After sunrise the situation changes as the land surface heats the air and vertical momentum transport is enhanced due to buoyancy. Therefore, during daytime the wind speeds near the surface are on average higher than at night. So the "W" shape in the bias might be caused by imperfect modelling of this effect by the NWP model.

However, for offshore sites or islands the atmospheric stability is typically different from onshore as the large heat capacity of the sea dampens temperature changes of the air due to solar irradiation. Hence, diurnal variations at sea are expected to be weaker compared with those over land but can be noticeable if, e.g. the continental coast is not too far away and the wind comes from that direction carrying properties of the land mass, as described by Lange [61]. This seems to be the case for Fehmarn, which is an island but still close to the German coast. For this site the variation between night and day values of the mean annual wind speed of both predicted and measured wind is less pronounced compared with that for inland sites. As Fig. 6.7 suggests this behaviour is covered with higher accuracy by the NWP model.

The "level" of the bias, i.e. the average bias over t_{pred}, and the sign are expected to depend on the local conditions. The results for the majority of sites in this investigation seem to indicate that the overall bias is mainly related to the terrain type. In flat terrain the bias is positive, which is equivalent to a general overestimation by the prediction, while in complex terrain an underestimation leads to mainly negative bias. This can be caused by the fact that the roughness length z_0 in the NWP model represents an average over a grid cell of the size $14 \times 14\ \mathrm{km}^2$. Thus, it might deviate from the optimal roughness length at the position of the site. An adequate choice of z_0 plays a vital role in the refinement procedure of the power prediction system as z_0 directly influences the logarithmic wind profile (3.16). Moreover, in complex terrain the wind turbines are systematically erected at "optimal" locations, i.e. on top of hills rather than in valleys. This might lead to the underestimation of these sites by the prediction because the NWP model includes the average orography over the grid cell into the roughness length z_0. Hence, the NWP model uses a meso-scale roughness length representing an area rather than a local roughness length.

The average fluctuations of the error around its mean are described by $\mathrm{sde} = \sigma(\epsilon)$ normalised by the mean wind speed in Fig. 6.8. Here the increase in the error with increasing prediction time occurs again with almost the same slope for all sites. Note that the curves for the different sites are closer together compared to the rmse in Fig. 6.6. These values directly provide the 68% confidence interval as the underlying error distributions of the wind speed prediction are normal, which was shown in Sect. 6.3. Hence, e.g. for the 6-h prediction at Rapshagen the sde allows for the statement: "With a probability of 68% the deviations of the relative forecast errors from the bias are less than 24% of the mean wind speed." This type of information is contained in Fig. 6.8 for the selected sites and all lead times.

Now sde is decomposed in sdbias, which is the difference between the standard deviations of predicted and measured time series, and disp, which is the contribution of the phase error to the rmse. It can be seen in Fig. 6.9 that the different sites have a rather different relative sdbias. While sites in flat terrain show a very small difference between the standard deviations of the prediction and the measurement, complex terrain sites provide a more substantial deficit in that respect ranging up to −20% of the mean wind speed at Altenbeken. There is no evident overall increase with t_{pred}. Similar to the bias of the mean values, the bias of the standard deviations does not scale with the mean wind speed.

Finally, the relative dispersion is presented in Fig. 6.10. For the different sites the relative dispersion is in a rather narrow range 0.23–0.27 for the 6-h prediction and 0.34–0.43 for the 48-h prediction. Again the increase with t_{pred} is obvious and closely similar to the behaviour of the relative sde in Fig. 6.8. From 6 to 48 h, the relative dispersion grows by about 0.13, which is of the order of 0.5 m/s for a site with 4 m/s average wind speed at 10-m height.

According to (6.3), disp is proportional to $\sqrt{\sigma(u_{\text{pred}}),\, \sigma(u_{\text{meas}})}$, which together with the ratios in Table 6.2 explain the rather universal behaviour of the relative dispersion for the different locations in Fig. 6.10. The differences that still remain are mainly due to different cross-correlation coefficients $r(u_{\text{pred}}, u_{\text{meas}})$. So, on the one hand the dispersion is universal for all sites in the sense that the influence of phase errors grows with prediction time at almost the same rate. Thus, all sites experience phase errors in a similar way. On the other hand disp has a local component as it is proportional to the amplitude variations of the prediction and measurement. The larger the amplitude variations, the larger the deviations that occur on average.

In Fig. 6.11 the complete decomposition of the rmse according to (6.3) is presented for one site (Hilkenbrook). At this site the bias is relatively prominent and introduces strong diurnal variations to the overall forecast error. If the bias is removed from the rmse the resulting standard deviation of error, sde, reveals the systematic increase in the phase error with the prediction horizon. As the sdbias is practically not relevant, the dispersion equals the sde. Hence, though the bias, i.e. the average amplitude errors, is considerable, the rmse is mainly dominated, by the phase errors.

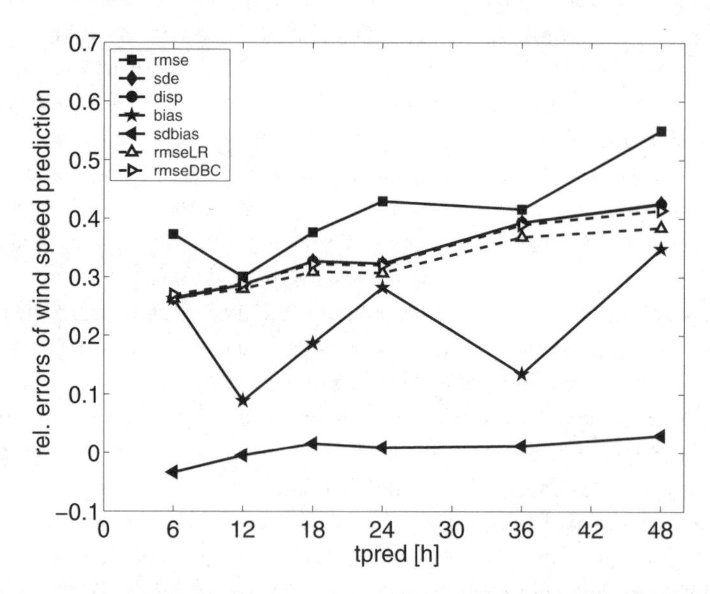

Fig. 6.11. Summary of the decomposition of the relative rmse of the wind speed prediction according to (6.3) versus forecast horizon for one site (Hilkenbrook). The dashed lines illustrate the rmse that can be expected if linear corrections are applied. rmse_{DBC} refers to the DBC correction (6.13), while rmse_{LR} (6.10) denotes the minimum rmse that can be expected by using a correction based on linear regression

The results of the two different linear correction schemes discussed in Sect. 6.4.2 are also illustrated in Fig. 6.11. The minimum rmse that can be achieved by linear regression (6.8) is shown by $\mathrm{rmse}_{\mathrm{LR}}$. It is, as expected, only slightly lower than the forecast error based on the DBC correction (6.11). In both cases the correction of the prediction leads to rather substantial improvements, in particular for the largest forecast horizon of 48 h. The relative improvement due to the LR correction expressed by the ratio (rmse–$\mathrm{rmse}_{\mathrm{LR}}$)/rmse is about 0.2 at this site. Compared to the sites in the present and previous investigations, e.g. by Giebel [35], this is around the maximum improvement that can typically be achieved for the wind speed prediction.

One advantage of the decomposition of the rmse as it is used here is that it indicates the expected benefits of linear correction schemes. The degree of error reduction that can be achieved by these methods strongly depends on the structure of the forecast error due to the different contributions of amplitude and phase errors which are separately shown by bias and dispersion. If the bias is considerable, say around ± 0.2 of the average wind speed, there is a good chance to eliminate this contribution to the overall error by linear transformations of the time series. However, if the bias and the sdbias are small, the forecast error is dominated by phase errors which resist simple correction efforts and have to be addressed in a situation-dependent approach.

6.6 Power Prediction Error for Single Sites

After the overall prediction accuracy of the wind speed prediction has been assessed in the previous section, the prediction of the power output is evaluated in the following. In order to compare different sites the assessment of the power prediction error also requires some kind of normalisation. In this investigation the annual mean of the measured power output is taken as reference value because it inherently considers both the machine type with its rated power and the typical wind speeds at the location. Table 6.3 shows details concerning the turbines that are installed at the six selected sites. The given mean values refer to the year 1996.

The relative standard deviation, $\sigma(P_{\mathrm{meas}})/\overline{P_{\mathrm{meas}}}$, of the measured power output in Table 6.3 is larger than the relative standard deviation, $\sigma(u_{\mathrm{meas}})/\overline{u_{\mathrm{meas}}}$, of the time series of the wind speed measurement by a factor 1.8–2.6. This factor is due to the power curve and can be regarded as the effective non-linearity factor that describes the scaling of variations in the wind speed due to the local slope of the power curve as already mentioned in Chap. 4. If $P(u)$ was proportional to u^3 and the variations in u were small compared with the mean value, the ratio between the relative standard deviations of power output and wind speed would be around 3, corresponding to the average relative derivative.

The power predictions used in this chapter were obtained by using the logarithmic wind profile (3.16) to transform the wind speed from the height given by the

Table 6.3. Rated power and annual mean values of measured power output of the wind turbines at the selected sites with corresponding standard deviation[a]

Site name	Rated power (kW)	Mean measured power output $\overline{P_{\text{meas}}}$ (kW)	Standard deviation of measured power output, $\sigma(P_{\text{meas}})$ (kW)	Relative standard deviation, $\sigma(P_{\text{meas}})/\overline{P_{\text{meas}}}$
Altenbeken	150	30.2	37.9	1.25
Fehmarn	200	59.8	59.5	0.99
Hilkenbrook	80	12.0	16.7	1.39
Rapshagen	300	48.8	58.8	1.20
Schuelp	250	46.0	59.6	1.30
Soelmnitz	200	25.4	35.6	1.40

[a] The relative standard deviation is between 0.99 and 1.40, i.e. it is larger than the corresponding wind speed result by a factor 1.8–2.6.

NWP model of the German Weather Service to the hub height of the wind turbines. This wind speed was inserted into the corresponding power curve. The local refinement and the correction for thermal stratification as described in Chap. 4 were not included in the prediction model at this stage in order to concentrate on the major effect of the power curve.

Analogous to the statistical analysis of the wind speed, Figs. 6.12–6.16 show the different relative error measures versus prediction time.

The relative rmse of the power prediction (Fig. 6.12) is in the range 0.5–1.1 for 6 h and 0.8–1.45 for 48 h. Hence, its overall level is about two times larger than the relative rmse of the wind speed.

In Fig. 6.13 the relative bias shows a rather large discrepancy between the different sites, from 0.8 to −0.5, being considerably more pronounced compared to wind speed (Fig. 6.7). The diurnal variations are still of the order of 0.1–0.2 but mostly inverse with maximum bias at noon rather than midnight. Thus, the characteristic "W" shape of the wind speed bias is lost (cf.Fig. 6.7).

The inversion of the diurnal variations compared to wind speed indicates that the difference between mean values at daytime and nighttime is overestimated by the power prediction system. This is mainly due to the fact that stability effects are not included in the transformation to hub height (3.16). As shown in Chap. 7, this can lead to an overestimation of the wind speed prediction, and hence the power prediction, in situations of unstable thermal stratification which typically occur during daytime. Another influence on the bias of the power prediction is expected by the non-linearity of the power curve as the difference between the mean values of prediction and measurement at noon grows larger than at midnight because the midday means have larger absolute values.

As before, the sign of the bias seems to be connected to the terrain type. While flat terrain sites have positive bias, those in more complex terrain provide negative

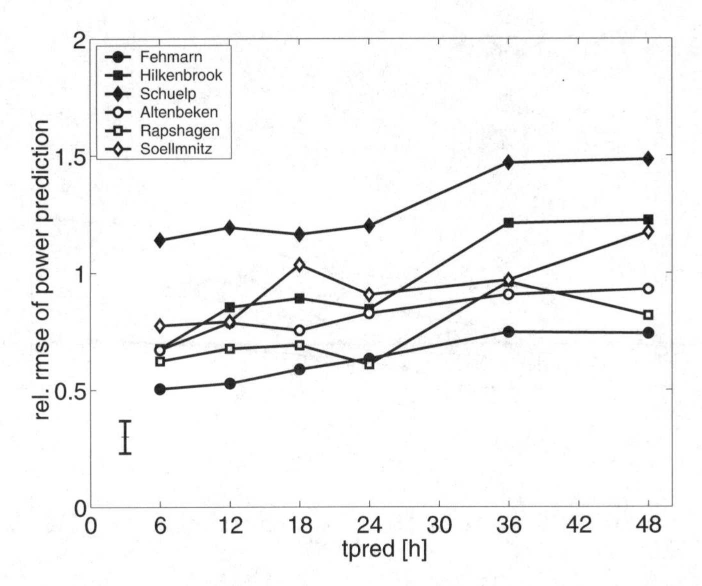

Fig. 6.12. Relative root mean square error (rmse) between predicted and measured power output versus prediction time for the selected sites in the year 1996. The rmse seems to increase with increasing forecast horizon. There is no apparent coherent behaviour among the different sites

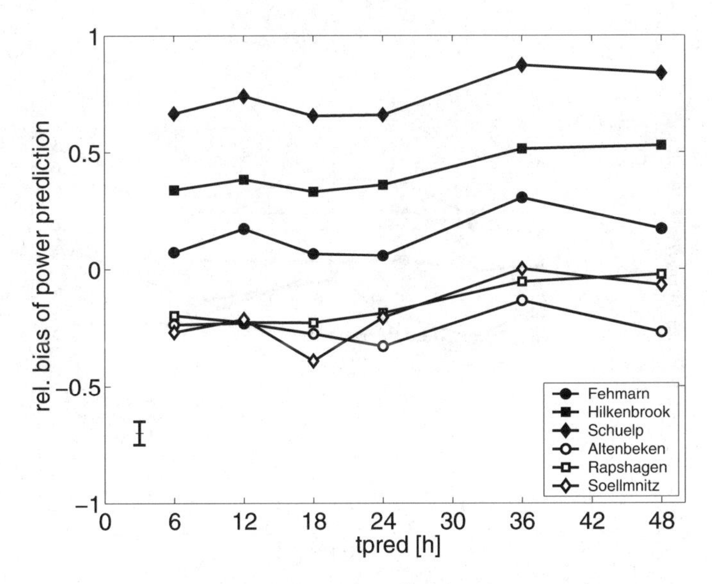

Fig. 6.13. Relative bias of the mean values of predicted and measured power output for the same sites as in Fig. 6.6. The diurnal variations are inverse compared with the wind speed bias in Fig. 6.7. Sites located in flat terrain have positive bias (*solid symbols*) while sites in complex terrain tend to have negative bias (*open symbols*)

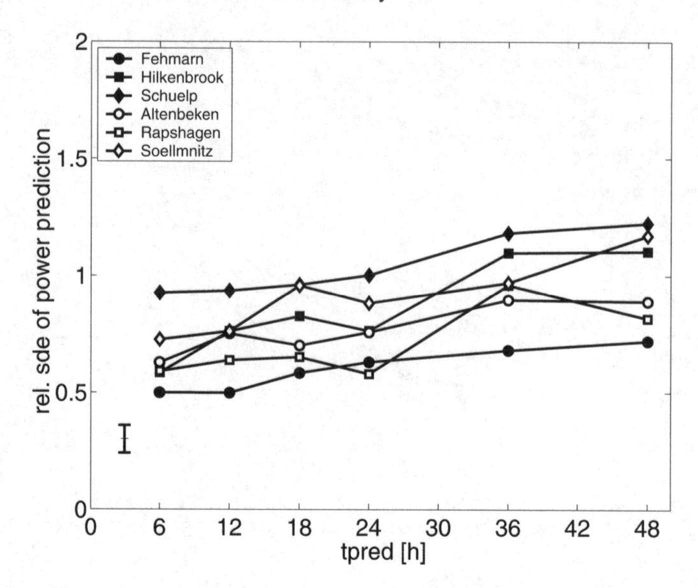

Fig. 6.14. Relative sde of the power prediction for the selected sites. Again the increase of sde with prediction time can be observed. The differences between rmse (Fig. 6.12) and sde of the power prediction is rather small for most sites. Hence, in contrast to the wind speed prediction (Figs. 6.6 and 6.8) the sde dominates the rmse with a smaller contribution of the bias

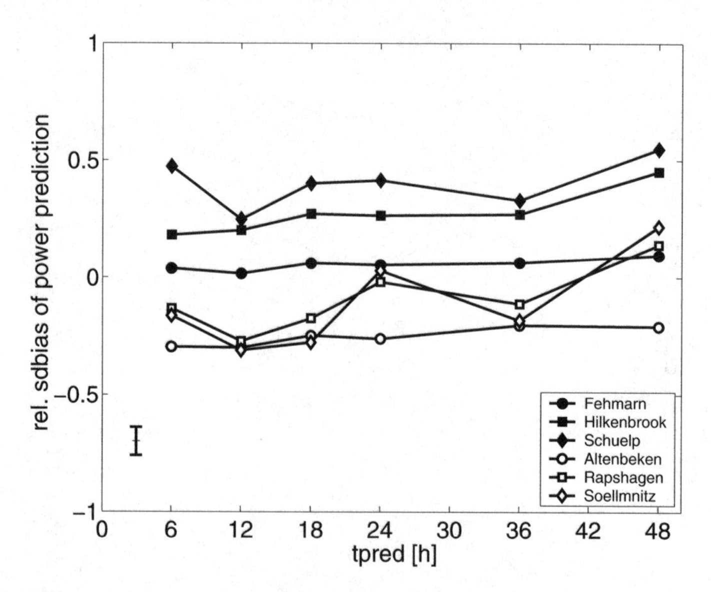

Fig. 6.15. Difference between standard deviations of predicted and measured power output (sdbias) normalised with mean power output for the selected sites versus prediction time. Like the sdbias of the wind speed in Fig. 6.9, the three flat terrain sites group together in the range of small sdbias

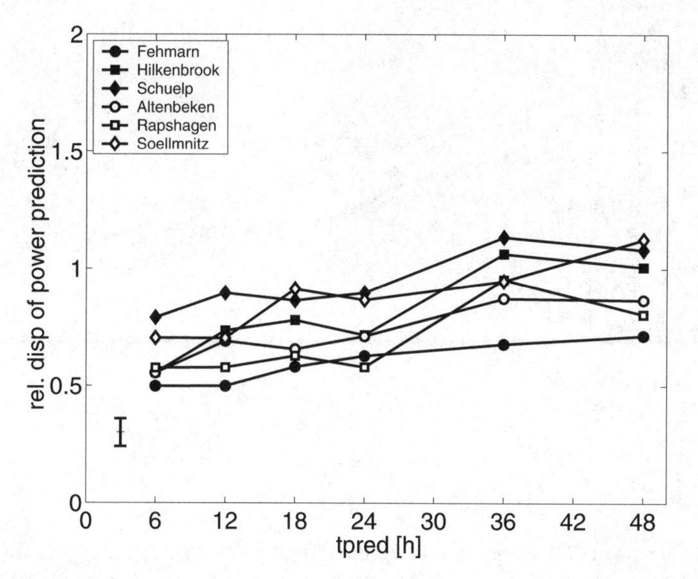

Fig. 6.16. The relative dispersion of the power prediction versus the forecast horizon is in a rather narrow range for the selected sites. It increases with prediction time with a similar rate at all sites indicating a systematic phase error that affects different sites in the same way

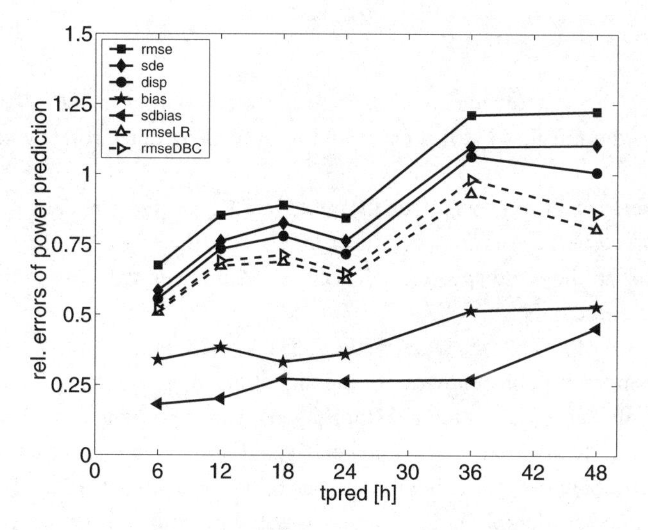

Fig. 6.17. Summary of the decomposition of the rmse of power prediction according to (6.3) versus prediction time for one site (Hilkenbrook). The dashed lines illustrate the rmse that can be expected if linear corrections are applied. $\mathrm{rmse}_{\mathrm{DBC}}$ refers to the DBC correction (6.13) while $\mathrm{rmse}_{\mathrm{LR}}$ (6.10) denotes the minimum rmse that can be expected by using linear regression

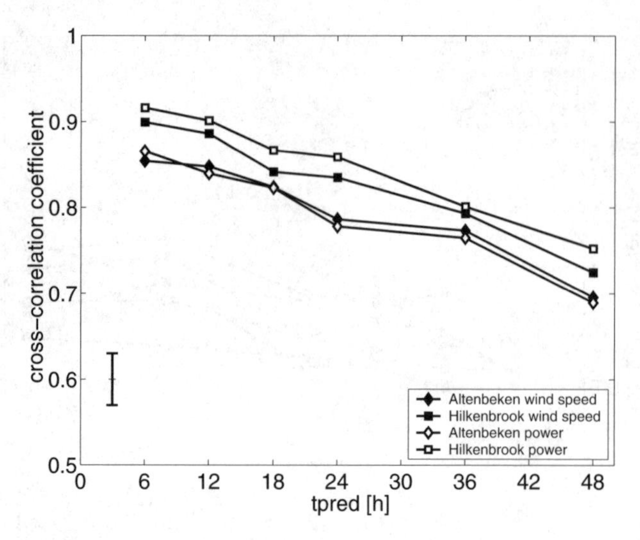

Fig. 6.18. Cross-correlation coefficients for the different prediction times of wind speed and power prediction at two different sites. Altenbeken has a comparatively low cross-correlation, while Hilkenbrook is among the sites with the highest values. At both sites there are only small differences between the two prediction types. This holds for all sites in this investigation. The cross-correlation suffers only minor changes by converting wind speed to power output using the *Previento* system

values. Thus, the general underestimation or overestimation, respectively, of sites according to the terrain is still present and enhanced by converting wind speed to power.

It can be seen in Fig. 6.14 that the relative sde, i.e. the standard deviation of the power prediction error, does not differ much from the relative rmse in Fig. 6.12. So in contrast to the wind speed prediction, sde dominates the rmse with a smaller contribution of the bias.

The relative sdbias of the power prediction (Fig. 6.15) is similar to the wind speed prediction concerning the behaviour of the individual sites. Again, being in complex terrain (Altenbeken, Soellmnitz and Rapshagen) leads to a general underestimation of the fluctuations of the measured power signal, while it is vice versa for flat terrain (Fehmarn, Schuelp and Hilkenbrook). Here the overestimation for Schuelp and Hilkenbrook is considerably larger compared with the wind speed results in Fig. 6.9.

Finally, regarding relative disp in Fig. 6.16, the different sites are rather similar. The overall level is larger than that of the wind speed by a factor of about? The differences among the sites are smaller compared with their rmse (Fig. 6.12), suggesting that the relative dispersion reveals some common statistical property.

A summary of the decomposition of the relative rmse of the power prediction at the site Hilkenbrook is shown in Fig. 6.17. Again, the minimum rmse that can be achieved by linear regression (6.8) and by the DBC correction (6.11) is illustrated by rmse_{LR} and rmse_{DBC}, respectively. In both cases the correction of the prediction would lead to rather substantial improvements for all forecast horizons. However, the overall error level that is not accessible with the linear correction methods remains larger than 0.5, which is still rather high.

It was said before that dispersion refers to phase errors by considering the cross-correlation between prediction and measurement and the individual amplitude variations of both time series. As Fig. 6.18 illustrates, for two different sites the cross-correlation coefficient has almost exactly the same values for wind speed and the power predictions. This holds for all sites that were investigated. Hence, the cross-correlation of the underlying wind speed prediction does not change much when it is converted to power output.

With the definition of the dispersion in (6.3) the doubling of the relative dispersion is caused by the standard deviations $\sigma(P_{\text{pred}})$ and $\sigma(P_{\text{meas}})$, which are relatively 2 to 2.5 times larger than $\sigma(u_{\text{pred}})$ and $\sigma(u_{\text{meas}})$ (cf. Tables 6.2 and 6.3). This factor can be considered as the effective non-linearity factor of the power curve. If the power curve followed u^3, this factor would be around 3.

The large standard deviations of the measured power time series, at a site are, of course, determined by the specific turbine and the local wind conditions. The prediction system is designed to match the measured time series as exactly as possible. Hence, it should provide the same mean and standard deviation as the "real" time series, together with a correlation close to 1. In this case the dispersion provides a lower boundary to the rmse for a given cross-correlation. This threshold is rather high in the case of the power prediction that already ranges up to 100% of the mean power output for the 48-h forecast. It is important to note that this level can only be reduced by improving the cross-correlation. Changes in bias and sdbias can optimise the power prediction only in a limited range.

In addition, Fig. 6.18 shows that the cross-correlation coefficients decay almost linearly with the forecast horizon. Note that each of these coefficients has been calculated separately with data points of one prediction time only. The decreasing cross-correlation coefficients indicate that as expected the mismatch between the prediction of the NWP model and local measurement grows systematically with the prediction horizon.

6.6.1 Comparison with Persistence

The DM prediction of wind speed is tested against persistence to compare the accuracy to a very simple reference system in order to get an impression about the different behaviour of the two. A persistence prediction is made by using the measurement value at 00 UTC as forecast value for the whole prediction horizon. Thus, the current value is assumed not to change in the near future, i.e. to persist over the

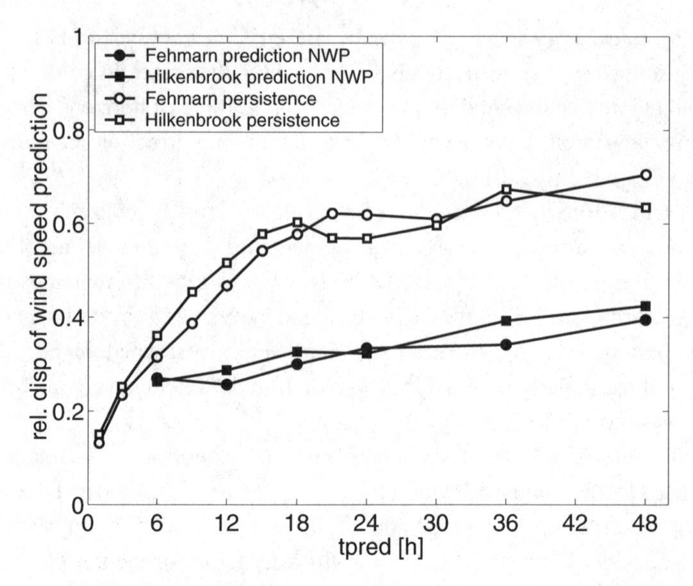

Fig. 6.19. Comparison between the relative dispersion of the persistence prediction of the wind speed, where the current measurement value is assumed not to change over the prediction horizon, and the NWP wind speed prediction for different prediction times. For lead times above about 6 h, the prediction generated by the NWP model performs significantly better than persistence

forecast interval. Persistence was chosen as reference because it inherently contains information about the actual meteorological situation at the site shortly before the prediction horizon begins.

It can be concluded from the relative dispersion in Fig. 6.19 that the prediction of a NWP model has a better performance than persistence for prediction times above 6 h. This is a well-known fact for numerical prediction systems, e.g. Landberg [58] and Giebel [35] showed that below 6 h, persistence is in fact better. The reason for this behaviour might be related to a relaxation process with decaying transients in the NWP model after the initial conditions are set and the run is started. The results for the power forecast are comparable.

Note that for forecast horizons larger than about 15 h, the climatological mean of the wind speed is a more suitable reference system than persistence. This is due to the fact that the error of the persistence forecast is by construction determined by the auto-correlation function of the wind speed, which decays approximately exponentially with a time constant of about 15 h, and, therefore, the persistence forecast is rather uncorrelated after this time period. Hence, a new reference model based on the superposition of persistence for short forecast horizons and the climatological mean for longer forecast horizons has been proposed by Nielsen et al. [81]. Though the use of this new reference model would reduce the difference between the errors of NWP forecast and the reference forecast by a few percent, the wind speed prediction of the NWP model is still significantly better.

6.7 Conclusion

The decomposition of the root mean square error (rmse) into different components provides considerable insight into the different origins of deviations between prediction and measurement. The contributions of the difference of the mean values (bias), the difference of the standard deviations (sdbias) and the dispersion (disp) to the rmse reveal some general statistical features of the errors that allow to distinguish local effects from global properties.

Concerning local effects, the results strongly suggest that the terrain type has a systematic influence on the bias and sdbias. For wind speed as well as power prediction, sites in flat terrain tend to have positive bias and small sdbias while sites in complex terrain are more inclined towards negative bias and sdbias. The reason for this is thought to originate from the fact that the NWP model uses a meso-scale roughness length which represents the typical surface roughness including orography within the grid cell and not the local situation at the position of the wind farm. In particular, in complex terrain the wind turbines are systematically erected at "optimal" locations, i.e. on top of hills rather than in valleys. This might lead to the general underestimation of these sites by the prediction. Hence, a general overestimation or underestimation, respectively, by the prediction can be attributed to the on-site conditions. However, in this investigation the terrain type was classified very roughly without quantitatively relating the behaviour of the bias and sdbias to orography.

Diurnal variations occur mainly in the bias indicating imperfect modelling of atmospheric stratification by the NWP model. With regard to the bias of the wind speed, most of the inland sites show a typical "W" shape where at midnight the relative bias is about 10–20% larger than at noon. At the island site this effect is weak. The bias of the power prediction is still of the order of 10% of the average annual power output but does not reproduce the "W" shape as the diurnal variations are reversed; this is mainly caused by neglecting the effects of thermal stratification in calculating the wind speed at hub height.

Regarding relative dispersion, in particular for the wind speed prediction, different sites show a very consistent behaviour indicating a general property that affects all locations in the same way. The linear increase of the dispersion with prediction horizon has comparable slopes for the majority of sites and describes the systematic growth of the average phase errors between prediction and measurement with increasing lead time.

The contribution of dispersion to the forecast error plays an important role. In contrast to the bias and the sdbias, it cannot simply be calibrated out by linear transformations of the complete time series as the dispersion involves the cross-correlation which is invariant under transformations of that kind. In this sense the dispersion provides an estimate of the lower boundary of the prediction error and limits the space for “simple” improvements.

Compared with the wind speed prediction, the relative dispersion of the power prediction is larger by a factor of 1.8–2.6. As the cross-correlation does not change by converting the wind speed prediction to power, this increase in dispersion is mainly caused by increased fluctuations, of the predicted and measured power time series, $\sigma(P_{\mathrm{pred}})$ and $\sigma(P_{\mathrm{meas}})$. Hence, the factor mainly represents the effective non-linearity of the power curve that unavoidably amplifies the wind speed fluctuations. Because larger amplitude variations are punished by a higher dispersion, the prediction error of the power prediction automatically increases compared with the wind speed prediction. Thus, the relatively large error of the power prediction system is not mainly caused by the additional errors associated with the modelling used in the power prediction system, e.g. local refinement. Of course, the procedure that calculates the predicted power has to be as precise as possible, but the crucial point is that the high error level of the power prediction more or less directly reflects the accuracy of the underlying wind speed prediction in combination with the influence of the non-linear power curve.

The lesson learned from this chapter is that converting a wind speed prediction to a power prediction fundamentally changes the statistical properties of the prediction error. Most of these changes can be blamed on the non-linearity of the power curve. The error of the power predictions has a non-Gaussian distribution, while the underlying wind speed errors are distributed normally in most cases. The prediction of the power output naturally has a larger error compared with the wind speed. Hence, the power prediction system primarily has to avoid additional errors to prevent things going from bad to worse. Correction schemes based on linear regression techniques that are applied globally to the time series are expected to be of little use to dramatically decrease the forecast error. One way out of this situation seems to focus on situation-based improvements of the wind speed prediction. This would require to work directly on the NWP model and enhance its performance with respect to an accurate modelling of the wind speed.

7

Correction of Wind Profiles Due to Thermal Stratification

Abstract. In this chapter the influence of thermal stratification on the wind profile is investigated in detail. The statistical distribution reveals that non-neutral situations occur rather often and on average the stratification is slightly stable. Classifying measured vertical wind profiles into different stability classes shows that absolute wind speeds and, in particular, the gradients strongly depend on the degree of stability of the lower boundary layer. Moreover, the Monin–Obukhov length L determined from atmospheric data according to the deBruin method proves to be robust enough to serve as a stability parameter in general situations despite the rather restrictive assumptions under which L is theoretically derived. Hence, using L together with empirical functions to correct the well-known logarithmic profile due to thermal stratification leads to major improvements in transferring the wind speed to hub height for heights up to 80 m. Though the prediction of the degree of thermal stratification by the numerical weather prediction system *Lokalmodell* of the German Weather Service has a rather low accuracy, the investigations for two wind farms show that taking thermal stratification into account significantly improves the wind power forecast.

7.1 Occurrence of Non-neutral Conditions

In order to estimate the influence of thermal stratification it is important to know how often different stability types occur in real atmospheric states. Figure 7.1 illustrates the frequency distribution of the Monin–Obukhov length L for a typical onshore site in flat terrain over 2 years. The stability parameter L was calculated according to the parametrisation scheme of deBruin et al. [16] based on measurements of wind speed and temperature at 2-m and 40-m height. Situations with stable stratification of the lower atmosphere are obviously more frequent than unstable situations. The maximum of the distribution indicates that most of the time the stratification is slightly stable.

Comparing the vertical wind profiles according to different stability classes illustrates the strong impact of thermal stratification on the profile that is theoretically

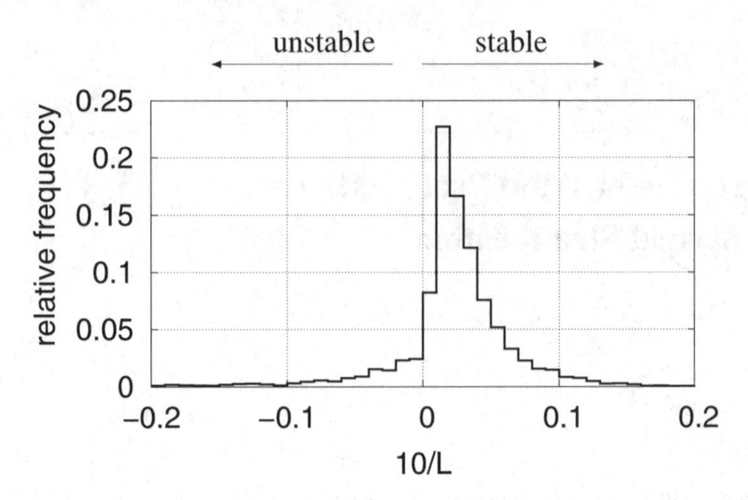

Fig. 7.1. Frequency distribution of thermal stratification determined by the stability parameter L. The maximum of the distribution is clearly in the range of stable stratification. Hence, at this onshore site most of the situations are slightly stable

derived in Chap. 3. Instead of the Monin–Obukhov length, which is estimated using the velocity gradient, the stability classes are now defined according to the gradient of potential temperature. In Table 7.1 the frequency of different stability classes defined by differences in potential temperature, $\Delta\Theta$ (3.18), at two heights (2 and 40 m) is shown.

The mean wind profiles corresponding to the stability classes in Table 7.1 are shown in Fig. 7.2 for unstable situations and in Fig. 7.3 for stable situations. Each of the mean profiles has been calculated by averaging over individual profiles of the same stability class.

Compared with the near-neutral case (class 0), the mean wind speeds in unstable situations are much smaller with smaller gradients. Hence, in the 30–200-m range the wind speed increases only little on average. This is due to the turbulent mixing in the lower part of the atmosphere evoked by buoyancy. As the difference between wind speeds at different heights is small, there is a strong coupling between them, and

Table 7.1. Frequency distribution of stability classes at Cabouw. The classification is based on the gradient of the potential temperature at the heights 2 and 40 m and ranges from "very unstable" (class −3) via "neutral" (class 0) to "very stable"

Class	−3	−2	−1	0	1	2	3	4
$\Delta\Theta$	< -1.4	$[-1.4, -0.8]$	$[-0.8, -0.3]$	$[-0.3, 0.3]$	[0.3,1.2]	[1.2,3]	[3,7.2]	>7.2
Frequency (%)	0.05	3.14	8.80	19.68	42.10	17.68	7.84	0.71

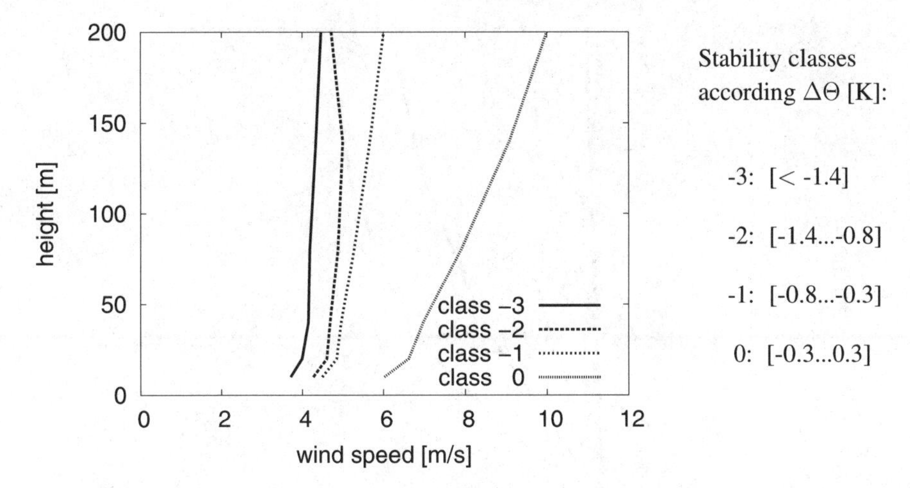

Fig. 7.2. Mean measured wind profiles for unstable stratification. The difference in the potential temperature at 2-m and 40-m height determines the stability class

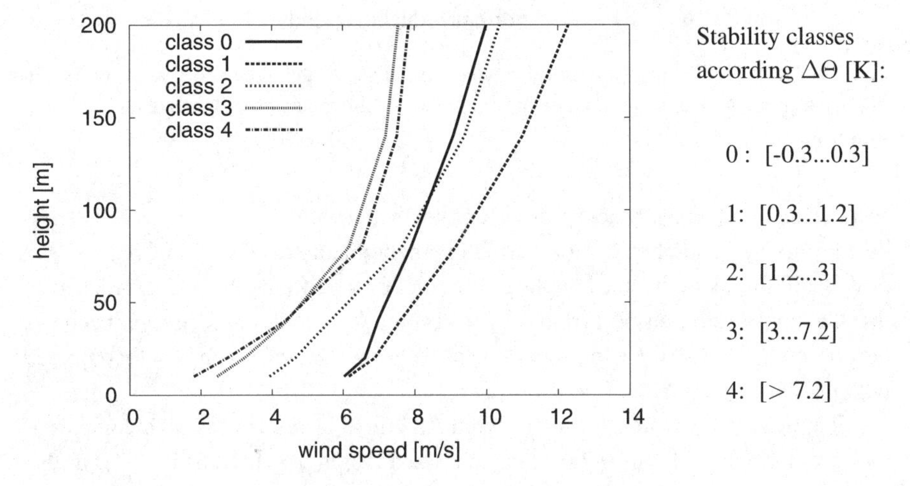

Fig. 7.3. Mean measured wind profiles for stable stratification. The difference in the potential temperature at 2-m and 40-m height determines the stability class

thus an effective downward transport of horizontal momentum. However, situations with unstable stratification do not occur very often. For the example site, only 12% (Table 7.1) of the atmospheric states are considered to be unstable, which agrees with other investigations, e.g. [105].

In contrast to this, the majority (approximately 68%) of situations are stable. It is important to note that these states are typically related to higher wind speeds at heights that are relevant for wind energy use as the gradients are typically considerably steeper compared with the near-neutral profile, in particular, for heights up to

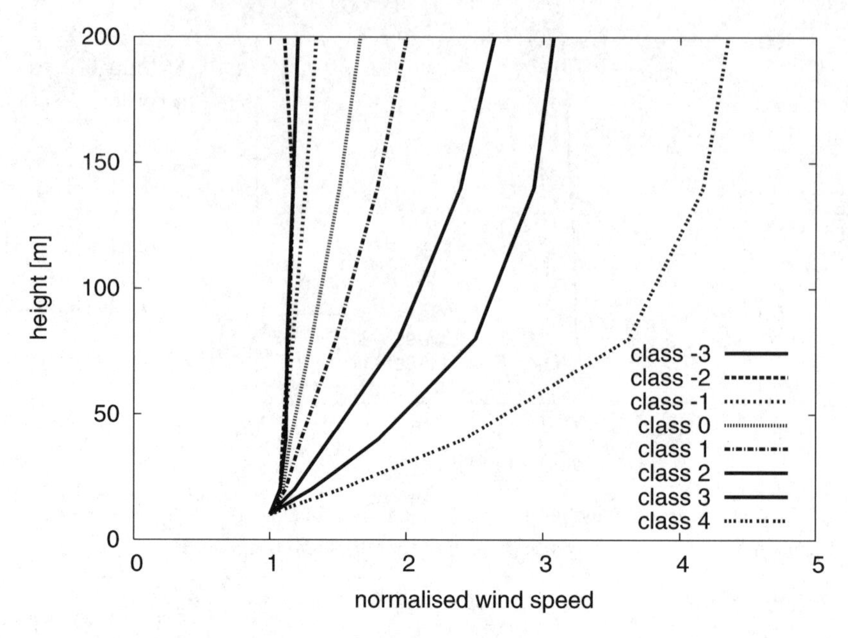

Fig. 7.4. Mean profiles for all stability classes normalised to the wind speed at 10-m height. The wind speed gradients up to a height of approximately 150 m strongly depend on thermal stratification

150 m (Fig. 7.3). The physical reason for this phenomenon is the suppression of turbulent, mixing in stable situations, which is theoretically discussed in Chap. 3. The flow is still turbulent, but the coupling between flow at different heights is reduced so that the momentum transport from the geostrophic wind to the surface layer becomes less effective. Hence, the wind speeds near the ground are relatively small compared with the wind speed in the free atmosphere.

The mean wind profiles of the different stability classes are easier to compare if they are normalised to the wind speed at a fixed height, e.g. 10 m (Fig. 7.4). In this representation it can be clearly seen that the slopes continuously decrease from stable to unstable situations. The unstable profiles with their flat gradients can clearly be separated from the stable profiles which are characterised by a pronounced increase of wind speed within the first 100 m from the ground.

For very stable stratification (class 4) the height of the surface layer is expected to be below 100 m (see e.g. [44]). This would be indicated by a decrease in wind speed above this height, which is clearly not detectable in the profiles in Fig. 7.4.

The mean wind profiles of the various stability classes are strongly different showing that thermal stratification has a significant impact on the shape of the profiles. At the site of Cabouw, which is comparable to typical flat-terrain sites in the north of Germany, about 80% of the situations are non-neutral. Hence, for wind

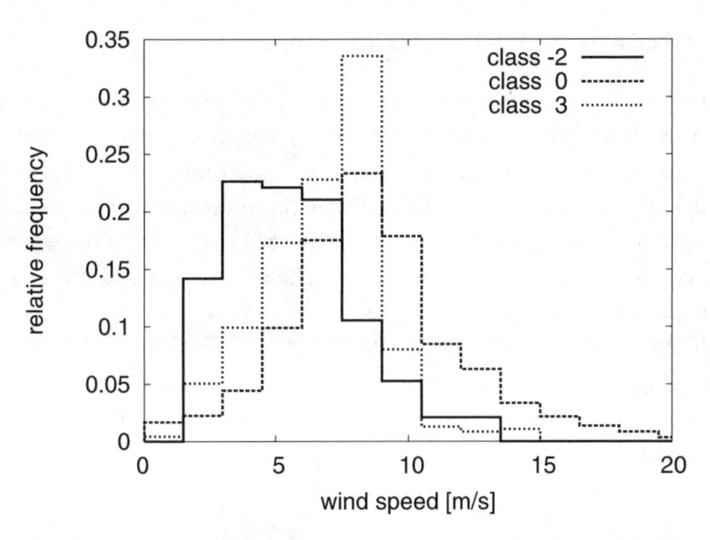

Fig. 7.5. Frequency distribution of wind speed at 80-m height for the stability classes −2, 0 and 3. In contrast to the wind speed at 10 m, stable situations are related to rather large average wind speeds at 80 m, and this relevant for wind turbines

power application the stratification of the atmosphere has to be considered, in particular, to determine the correct wind speed at the hub height of the wind turbines.

Finally, Fig. 7.5 shows the statistical distributions of the wind speed at 80 m for unstable, neutral and stable stratification. Wind speeds greater than 10 m/s primarily occur in neutral situations. It can also be seen that stable situations which have rather low wind speeds at 10 m can have considerable wind speeds at 80 m. This is, of course, important to know for wind power predictions. If low wind speeds are predicted at 10 m the relevant wind speed at hub height (e.g. about 80 m) might nevertheless be very significant in stable stratifications.

7.2 Application of Monin–Obukhov Theory

The Monin–Obukhov theory is strictly valid only for a very limited range of real atmospheric conditions, in particular for stationary situations. In this section the investigation focus on the applicability of the Monin–Obukhov approach in situations where the assumptions under which it is derived are not valid. Moreover, the influence of typical variables such as the magnitude of wind speeds and temperatures or the roughness length on the thermal correction is evaluated. The aim is to assess the benefits of using the correction scheme for practical purposes.

7.2.1 Correction of the Wind Profile up to 40 m

In the lower part of the wind profile up to around 40 m the gradients are generally large and strongly depend on the thermal stratification, this can be clearly seen in Fig. 7.4. Hence, at these heights a correction of the profile according to the thermal stratification is expected to significantly improve the transformation of the wind speed from one height to another. Hence, the thermal correction of the wind speed at 40-m height based on the wind speed at 10-m height is calculated using the neutral profile on the one hand and the corrected profile on the other.

As discussed in Sect. 3.4, thermal stratification is implemented as a correction function $\psi(z/L)$ to the neutral logarithmic profile. To obtain the thermally corrected wind speed u_{z_2} at height z_2 based on the wind speed u_{z_1} at z_1, the equation

$$u_{z_2} = u_{z_1} \frac{\ln\left(\frac{z_2}{z_0}\right) - \psi\left(\frac{z_2}{L}\right)}{\ln\left(\frac{z_1}{z_0}\right) - \psi\left(\frac{z_1}{L}\right)} \tag{7.1}$$

is used. The stability parameter L is determined according to the scheme by deBruin [16] using the temperature differences at 2-m and 40-m height. To exclude roughness effects, only data of one homogeneous sector (180°) are considered.

Figure 7.6 shows the ratio between calculated wind speed at 40 m and measured wind speed at the same height. The ratio is plotted against the thermal stratification of the atmosphere expressed by z_1/L, where, as before, $z_1/L = 0$ refers to a neutral stratification, $z_1/L < 0$ to an unstable situation and $z_1/L > 0$ to a stable state. If only a neutral profile is used (Fig. 7.6 (left)), i.e. $\psi = 0$, the wind speed is clearly overestimated in unstable situations while it is underestimated in stable conditions. This is perfectly understandable from the considerations in Sect. 7.1, where the thermally induced deviations from the neutral profile are discussed. Thus, the

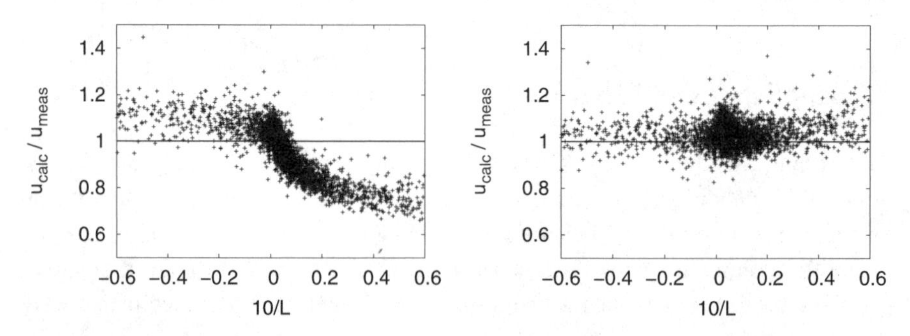

Fig. 7.6. Ratio of calculated and measured wind speed at 40-m height versus stability parameter $10/L$. The wind speed is calculated at 10 m height using the neutral profile (*left*) and the thermally corrected profile according to (7.1) (*right*). L is determined by the temperature difference at 2-m and 40 m height

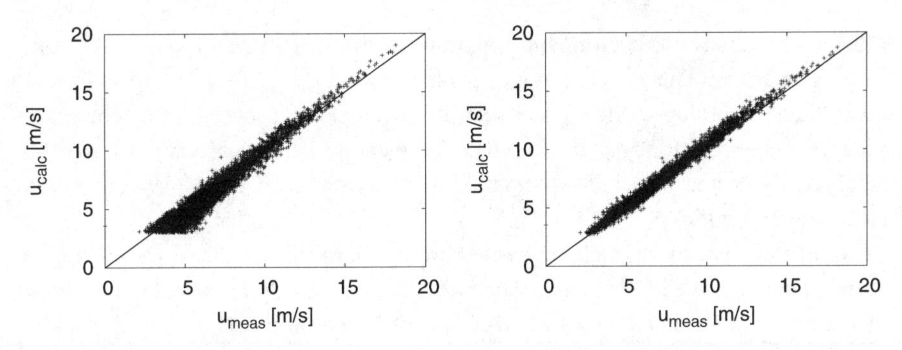

Fig. 7.7. Calculated versus measured wind speed at 40-m height. As in Fig. 7.6 the calculation is based on the wind speed at 10 m using the neutral wind profile (*left*) and the thermally corrected profile (*right*)

data in this case show that applying the neutral profile in every situation leads to considerable errors (around 20%) in calculating the wind speed at different heights. Fortunately, the right thermal correction according to (7.1) significantly improves the results, as shown in Fig. 7.6 (right). The systematic deviations from the optimal value are mostly eliminated, leaving only a slight overestimation of the wind speed for $10/L > 0.3$.

The thermal correction is carried out for each situation individually according to the specific value of L. Hence, not only general trends due to systematic over- or underestimations are corrected, but the scatter around the desired value is reduced. In Fig. 7.7 the scatter plots are shown, where the calculated wind speed at 40-m height is plotted against the corresponding measurement. It is clearly visible that the variations around the diagonal are considerably larger if the neutral profile is used (Fig. 7.7 (left)) compared with the thermally corrected profile (Fig. 7.7 (right)).

The general improvement by using the thermal correction can be expressed in terms of statistical error measures (c.f. Chap. 6).

Table 7.2 reveals that the bias changes sign and is slightly positive after the correction. But as the absolute value of the bias is rather small, this is only a minor change. What is more important is the increase in the cross-correlation from 0.97 to 0.99, as this improvement indicates the benefitial use of a situation-based correction

Table 7.2. Comparison of statistical errors in calculating the wind speed at 40 m height based on the wind speed at 10 m using the neutral profile and the thermally corrected profile

	$\text{bias}_{u_{\text{calc}},u_{\text{meas}}}$	$r_{u_{\text{calc}},u_{\text{meas}}}$	$\sigma(u_{\text{calc}} - u_{\text{meas}})$
Neutral	-0.272 ± 0.004	0.974	0.662 ± 0.007
Corrected	0.229 ± 0.003	0.992	0.348 ± 0.003

scheme. The cross-correlation is invariant under linear transformations of the complete time series. Therefore, a significant increase in $r_{u_{\mathrm{calc}},u_{\mathrm{meas}}}$ due to individual corrections according to the current stability parameter L shows that the correction scheme indeed has a physical meaning. The increased cross-correlation leads to a significantly decreased standard deviation of the error between calculated and measured wind speed.

Hence, the use of thermally corrected profiles according to the Monin–Obukhov theory instead of the neutral logarithmic profile considerably improves the accuracy of determining the wind speed at 40-m height based on the 10-m speed. Though the Monin–Obukhov approach is strictly valid only in stationary situations, the scheme used here, including the stability parameter L calculated according to deBruin et al. [16] together with the standard empirical correction functions, appears to be robust enough to produce reasonable results under various meteorological conditions.

7.2.2 Correction of Wind Profile Above 40 m

Having shown that including the effect of thermal stratification into the calculation of the wind speed at 40-m height leads to a better accordance with measurements, our focus is now on the greater heights which are more relevant for wind power applications. Naturally, the uncertainty of correcting the wind profile increases with height, and hence the accuracy of the correction scheme is evaluated for these cases.

In the following, the scheme that has been successfully used in the last section is applied to calculate the wind speed at 80 m and 140 m. The 80-m wind is determined based on the wind at 10-m and 40-m height, respectively, while the 140-m wind is based on the 40-m wind only. The Monin–Obukhov length L is again generated from the temperature at 2 m and 40 m.

Figure 7.8 shows, analogous to Fig. 7.6, the ratio between calculated and measured wind speeds for the different heights. As before, the wind speeds are overestimated in unstable situations and underestimated in stable stratification if the neutral profile is used to transfer the wind speed from one height to another. In particular, for the large height differences 10–80 m and 40–140 m, the underestimation in stable conditions is rather pronounced and reaches up to 40%. If thermal stratification is included the systematic deviations are strongly reduced and the distribution of the speed ratio is around 1 indicating that the correction scheme is also applicable for greater heights.

However, in statistical terms the achieved improvements differ for different heights. Figure 7.9 shows the error measures of various combinations of heights together with the 10–40-m results from the last section for comparison. Using thermally corrected profiles instead of the logarithmic profile reduces the bias considerably, in particular for large height differences; this is mainly due to eliminating

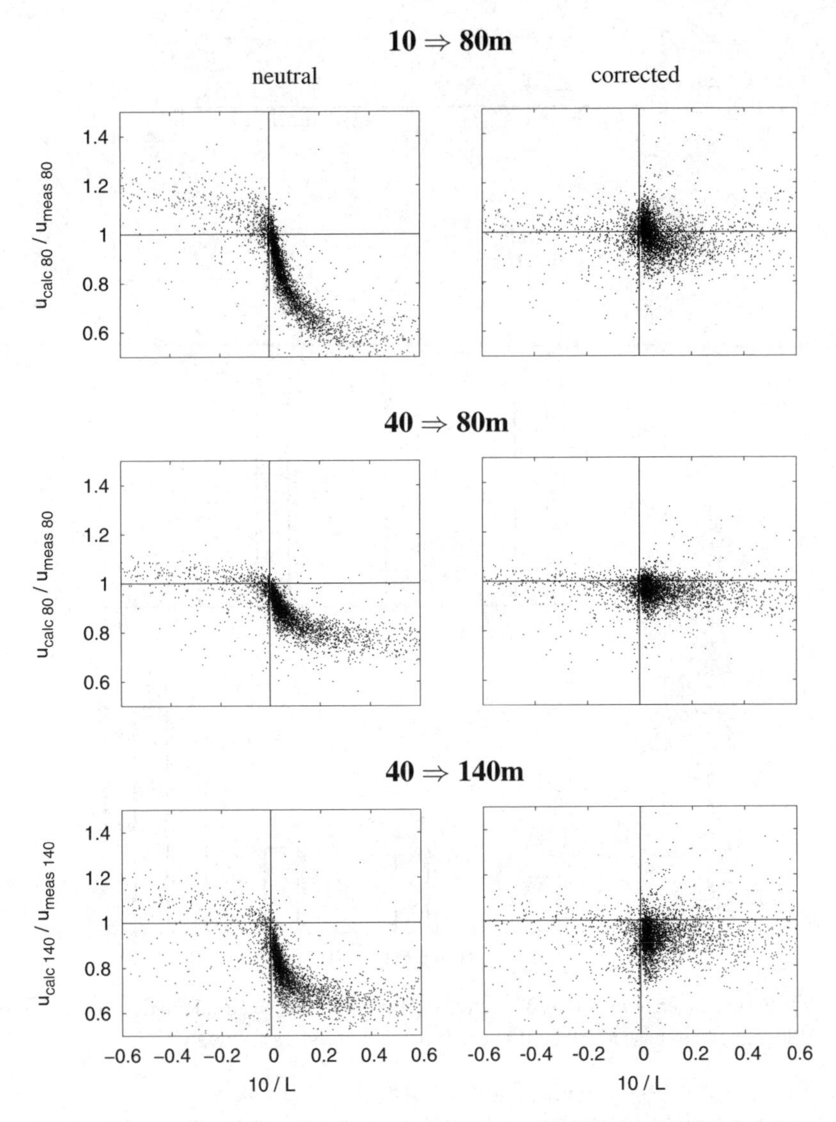

Fig. 7.8. Ratio between calculated and measured wind speed at 80-m and 140-m height versus stability parameter $10/L$. The calculated wind speed is determined based on the 10-m or 40-m wind speed, respectively, using the neutral profile (*left*) and under consideration of the thermal stratification (*right*)

the general underestimation of the wind speed in stable cases. Moreover, the cross-correlation increases for heights up to 80 m. In particular, for the 10–80-m extrapolation of the wind speed the cross-correlation rises dramatically from 0.93 to 0.98. Thus, if thermal stratification is considered the wind at 10 m contains as much information about the wind speed at 80-m height as the wind at 40 m does, and both

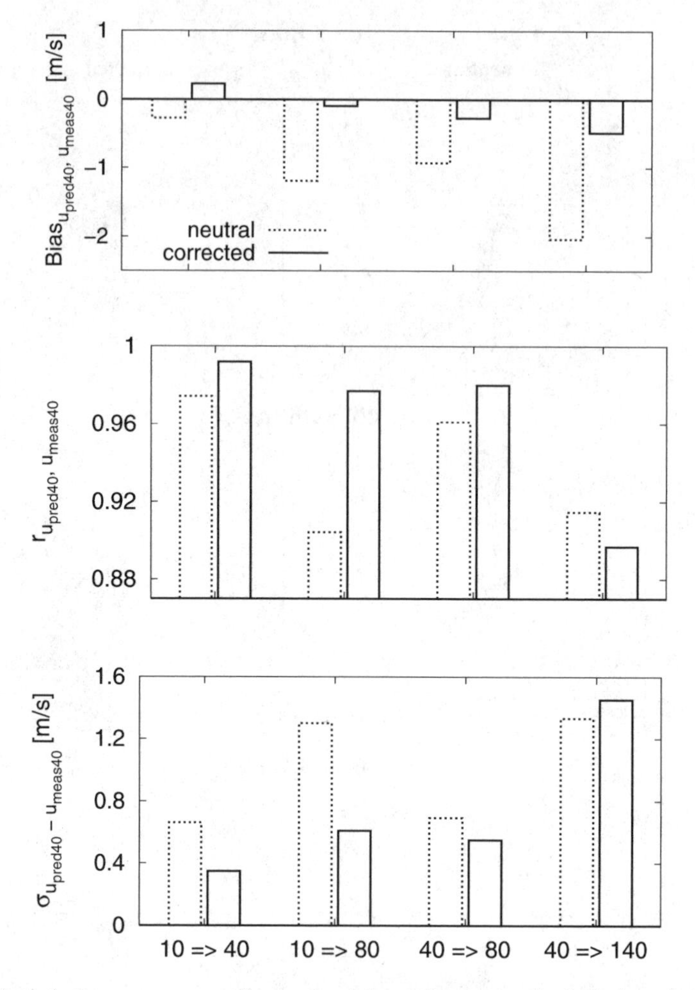

Fig. 7.9. Statistical error measures of neutral and thermally corrected calculation of wind speed at 40, 80 and 140 m based on the 10-m or 40-m wind speed, respectively

starting heights lead to the same cross-correlation. The mean square error σ also has the same level in both cases, so that starting from 10 m or 40 m does not make a big difference as long as thermal effects are included.

In contrast to this, the correction scheme is not successful for the wind at 140-m height. The improvement in the bias cannot compensate for the decrease in cross-correlation, so that the standard deviation of the error slightly increases. The results suggest that 140 m is beyond the surface layer, so that the logarithmic profile is no longer valid at this height (see Chap. 3) and, hence, the thermal correction of this profile also fails.

Generally, the correction scheme works well to find the wind speed up to heights of about 80 m under consideration of the thermal stratification where, surprisingly, the accuracy of the calculated wind speed does not depend on the starting height. As expected, the validity of the logarithmic profile and the Monin–Obukhov theory is limited and no longer applicable to heights around 140 m.

7.2.3 Influence of Roughness on Stability Correction

Commonly, the local roughness length z_0 is a subjectively determined external parameter. In order to estimate the influence of z_0 on the thermal correction the calculations of Sect. 7.2.1, i.e. from 10 to 40 m, are carried out with various roughness lengths. A recalculation is necessary because the influence of the roughness length on the thermal correction cannot easily be concluded from (7.1) as z_0 is also used to determine L. The range of z_0 at the investigated site would be 0.001–0.03 m. Note that z_0 is determined per sector.

The results for the pure logarithmic and the corrected profiles are shown in Fig. 7.10 in terms of the well-known error measures, where z_0 has been varied over three orders of magnitude from 0.0001 to 0.1 m. The bias strongly depends on the roughness length in both cases. This is because the choice of z_0 determines the vertical profile and influences how much the calculated wind speed systematically differs from the measured wind speed. Nevertheless, the difference between both methods is not much, so that the benefit of using the stability correction is virtually independent.

The same holds for the cross-correlation, which is only marginally influenced by the roughness length. If the neutral profile is used the cross-correlation does not change at all, as according to (7.1) the calculated wind speed is effectively multiplied by a roughness-dependent constant, i.e. $u_{\mathrm{calc}} = a(z_0)\, u_{10}$, which does not affect the correlation. The standard deviation of error, σ, depends on z_0 and profoundly increases for $z_0 > 0.01$ for the neutral and the corrected profile, but again the difference changes only slightly.

Hence, though the roughness length has a major impact on the magnitude of the calculated wind speed at a certain height, the advantage of using the thermally corrected profile instead of the neutral one remains basically the same.

7.2.4 Dependence of Stability Correction on Temperature Differences

The Monin–Obukhov length L depends on the temperature gradient of the lower boundary layer. However, to determine the value of L a good choice of the involved parameters is necessary, and therefore the following question arises: which part of the temperature profile is most adequate to describe the influence of thermal stratification on the vertical wind profile? In order to find the optimal parameters the investigation

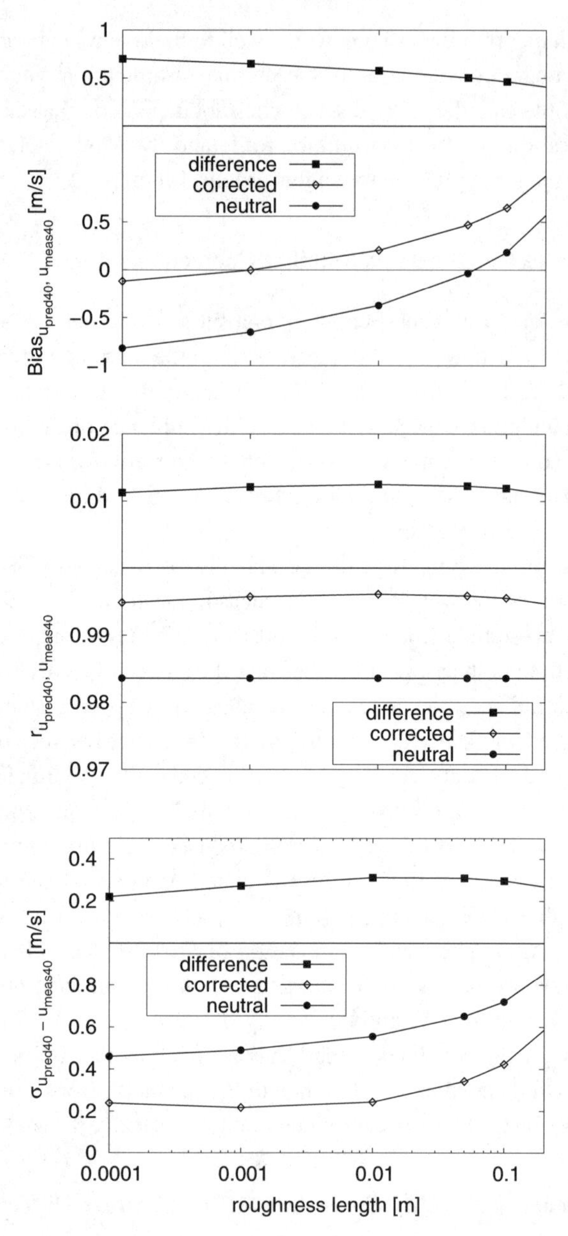

Fig. 7.10. Sensitivity of calculated wind speed at 40-m height depending on roughness length in terms of error measures. The bias (*top*), cross-correlation (*middle*) and standard deviation of error (*bottom*) are shown for calculations with neutral and thermally corrected profile. In the upper part of each figure the difference between both methods is shown

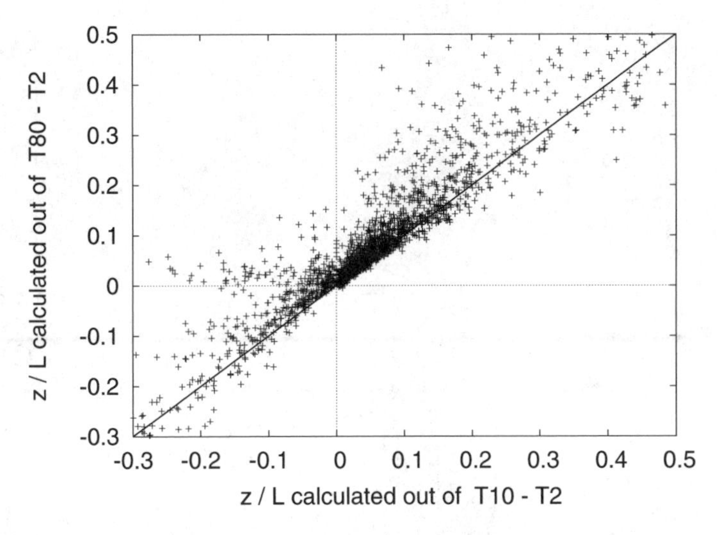

Fig. 7.11. Comparison of stability parameter z/L where the Monin–Obukhov length L is determined from the 2–10-m (*abscissa*) and 2–80-m (*ordinate*) temperature difference

in Sect. 7.2.1 is repeated using temperature differences from various combinations of heights.

The determination of L from different parts of the temperature profile is expected to lead to different results because the temperature difference that is used can be considered as a bulk gradient that contains information concerning the stability of the atmosphere over the chosen height difference.

In Fig. 7.11 two ways of calculating the Monin–Obukhov length based on rather extreme cases, namely the temperature difference at 2 m and 10 m, i.e. close to the surface, and the temperature difference at 2 m and 80 m, i.e. covering the complete lower boundary layer, are compared. For stable situations where both values of L are positive, the stability parameter originating from the 2–80-m gradient gives a smaller L and, hence, a larger z/L. In terms of the wind profile this means a larger correction. For unstable stratification, where L according to the difference between 2 m and 10 m is negative, it is vice versa and the 2–80-m temperature gradient leads to larger L. Moreover, a small set of situations is classified differently by both methods with L based on the 2–10-m differences being negative and that based on the 2–80-m differences being positive. These situations can typically be found in the early morning hours where the air near the surface is already heated from the ground and is locally unstable, while the stratification of the layers above is still stable.

Hence, the choice of the heights which are used to actually determine the temperature gradient has an impact on the value of the Monin–Obukhov length L. However, these slight differences do not have a major impact on the accuracy of the stability correction. The resulting error between calculated and measured wind speed

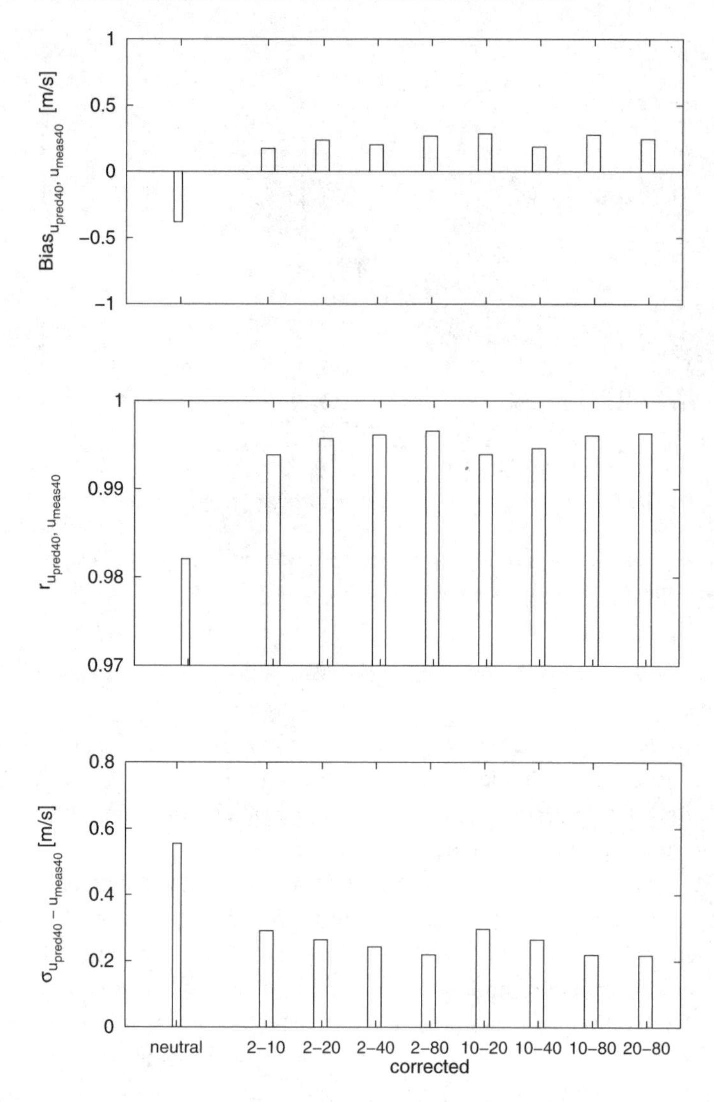

Fig. 7.12. Influence of different pairs of heights to determine the temperature difference on the calculated wind speed at 40 m. The notation 2–10 means that temperature measurements at 2-m and 10-m height are used to obtain L; similar notation is used for other combinations. For comparison the results for the neutral profile are given. The error measures used are bias (*top*), cross-correlation (*middle*) and standard deviation of error (*bottom*)

virtually does not depend on the choice of the temperature gradient. In Fig. 7.12 the usual error measures of the corrected wind speed at 40-m height are shown, where L has been calculated based on eight pairs of temperature differences. It can be concluded that all thermal corrections significantly improve the result compared with the neutral profile. Moreover, temperature gradients that involve the height of 80 m lead to a slightly better cross-correlation and lower standard deviation of error than gradients built from lower heights only.

Thus, a parametrisation of the thermal stratification can successfully be carried out based on different parts of the temperature gradient. To obtain best results the temperature gradient between around 20 m and 80 m seems to be most appropriate.

7.3 Prediction of Thermal Stratification

In order to use the stability correction of the wind profile for wind power forecasts, the predictability of thermal stratification by the numerical weather prediction (NWP) system has to be checked. For this purpose the accuracy of the NWP system in predicting temperature differences is assessed.

The NWP data used here comes from the *Lokalmodell* (LM) of the German Weather Service. Being a non-hydrostatic model the LM is in principle capable of describing local thermal effects. The predicted temperature difference between 2 m and 33 m provided by LM is compared with the measured difference between 2 m and 20 m. The comparison is helpful despite the deviating upper heights because large temperature gradients occur in the lower part of the profile between 2 m and 10 m.

Whether a situation is stable or unstable is quite accurately predicted by the NWP system. This is illustrated in Fig. 7.13, where the frequency of corresponding signs between the predicted and the measured temperature differences is shown. The average quality of the prediction varies over the day. During nights 95% of the situations are correctly classified. In this time interval, unstable stratification is very seldom,

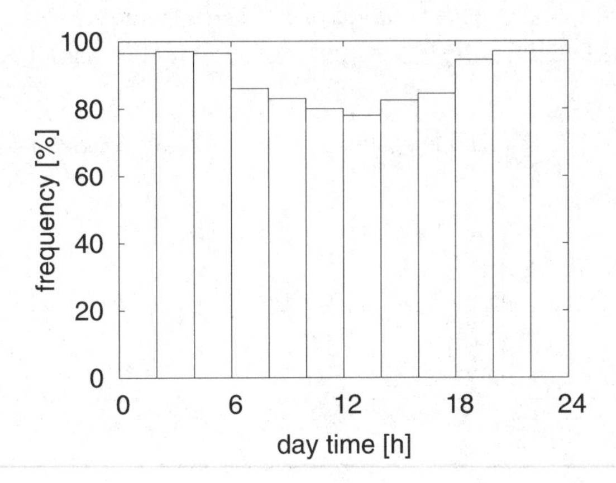

Fig. 7.13. Frequency of correspondence between the signs of predicted and measured thermal stratification

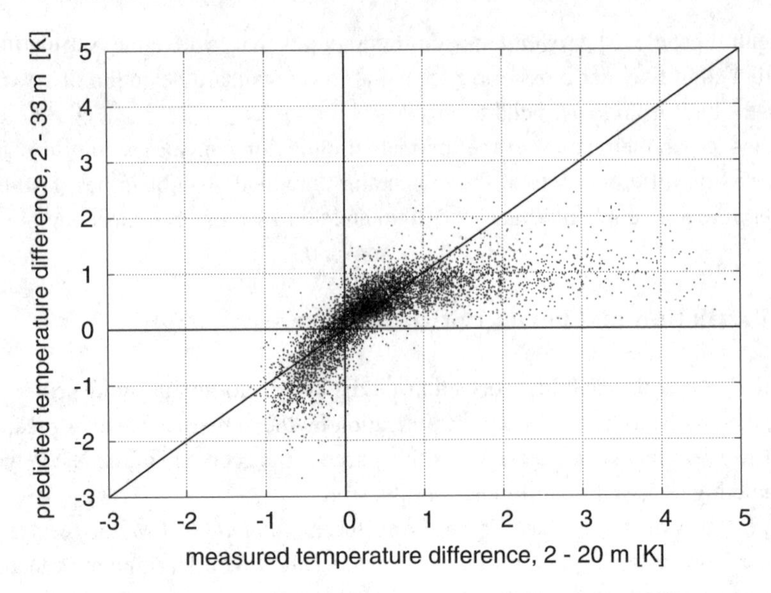

Fig. 7.14. Scatter plot of predicted versus measured temperature differences

and this simplifies the forecast. The rate of correspondance drops to around 80% during daytime, indicating that unstable conditions are harder to predict.

However, a more detailed assessment reveals that the prediction accuracy of the temperature differences is not satisfying in all cases. The scatter plot in Fig. 7.14 directly compares predicted and measured temperature differences and shows a good agreement only in the range $\Theta = -0.3$ K to $\theta = 1$ K. For larger gradients profound deviations occur; in particular, the measured positive differences are considerably larger than the predicted ones. The maximum predicted difference is 2.5 K, while the maximum measured difference is larger than 4 K. In contrast to this, strong negative gradients are predicted that have not been measured in this magnitude.

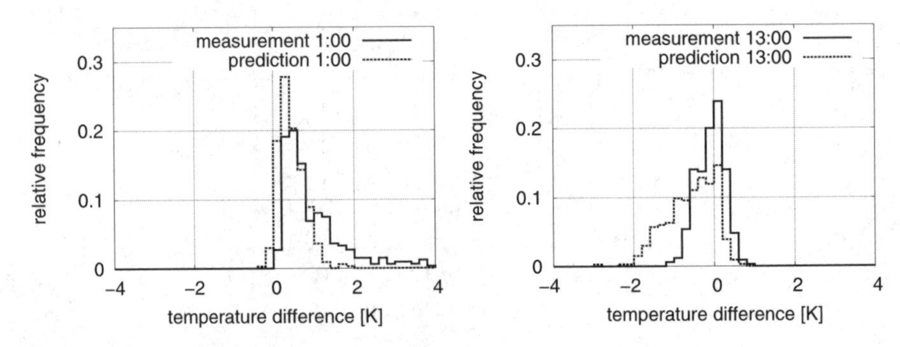

Fig. 7.15. Comparison of the frequency distributions of predicted and measured temperature differences at 1:00 CET (*left*) and 13:00 CET (*right*)

The frequency of these profound deviations can be evaluated by considering the statistical distributions of measured and predicted temperature differences. In the middle of the night, e.g. 1:00 CET, the majority of the situations are stable, and this is basically reflected by the prediction system (Fig. 7.15 (left)) except for the large temperature gradients. This changes completely after sunrise, so that at noon, i.e. 13:00 CET, the stratification is unstable in most cases (Fig. 7.15 (right)). Again the measured and predicted distributions do not match for large gradients. But at 7:00 CET and 16:00 CET both distributions are very similar (not shown), indicating that the difficulty in predicting large gradients are limited to rather extreme conditions, either stable or unstable.

The error measures bias, cross-correlation and standard deviation of error, sde, are used to quantitatively assess the prediction accuracy of the temperature difference forecast. The results are shown in Fig. 7.16. In particular, the cross-correlation varies dramatically over the day. While it is on an acceptable level of 0.75 during the daytime hours, it drops down to 0.5 during the night. This reflects the fact that the prediction fails for very stable situations with large positive temperature gradients, as already indicated by the scatter plot in Fig. 7.14. Consequently, the sde of the temperature prediction also rises for the nighttime hours.

Hence, using the predicted temperature differences of the NWP model rather accurately provides the stratification type in terms of the sign of the gradient, but the prediction of the absolute value of the temperature differences to determine the degree of stability appears to be rather inaccurate. In particular, a situation with large temperature gradients poses a problem to the NWP forecast. This might be due to the coarse vertical resolution of the NWP model.

7.4 Verification of the Stability Correction Scheme

In the following, the correction scheme to account for the effects of thermal stratification is applied to produce a wind power prediction for two wind farms and the results are compared with the real power output. For this purpose the scheme that has been discussed in the previous sections is implemented into the prediction system *Previento* (see Chap. 4). Naturally, the error of the wind power prediction for the test sites is accumulated over all effects in the model chain, particularly due to the initial uncertainty in the wind speed prediction of the NWP model, the transformation to hub height, the effect of the power curve and the shadowing effect within the wind farm. Hence, the question is whether the stability correction leads to an improvement in the final power prediction or is covered by the other various sources of error.

The wind power prediction is based on the wind speed prediction at 10-m height of the *Lokalmodell*. The wind speed at hub height is calculated under consideration

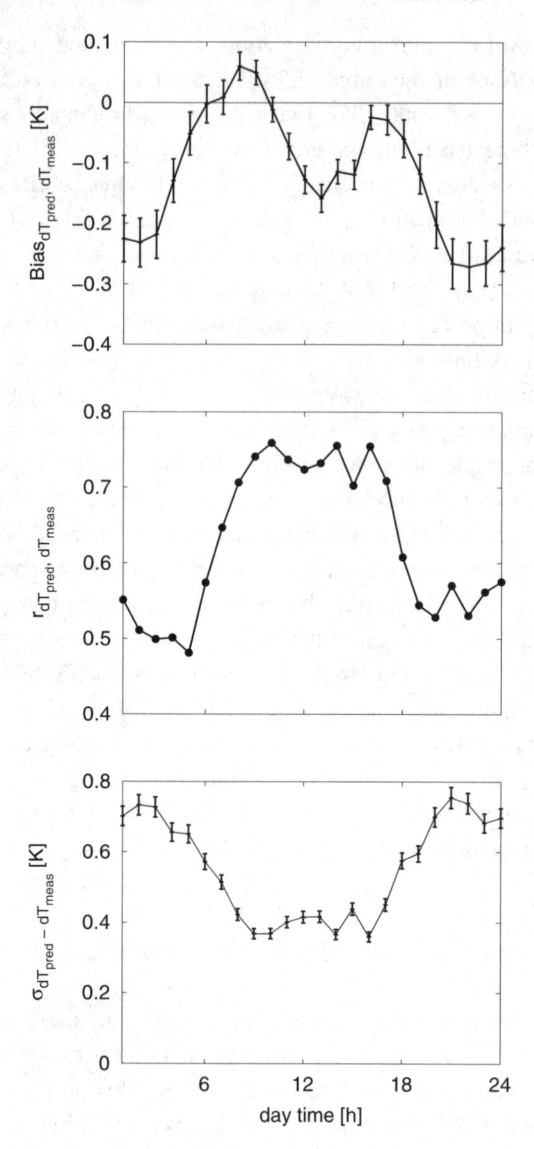

Fig. 7.16. Accuracy of temperature difference prediction by Lokalmodell over the day. The bias (*top*), cross-correlation (*middle*) and standard deviation of the error (*bottom*) are shown. The error bars illustrate the uncertainty of determining the error values

of a direction-dependent roughness description (see Sect. 4.3). To determine the Monin–Obukhov length L the predicted difference between the temperatures at 2 m and 33 m is used. The two test sites Neuenkirchen and Hengsterholz are located in northern Germany in flat terrain, so that orographic effects are not relevant.

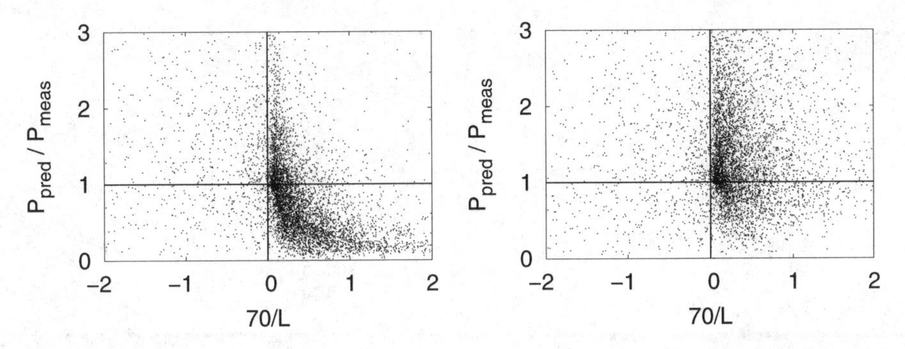

Fig. 7.17. Ratio between measured and predicted power output versus z/L for wind farm Neuenkirchen using neutral profile (*left*) and with stability correction (*right*)

Neuenkirchen is nearer to the coast, while Hengsterholz is a typical inland site. Certified power curves are used to transfer wind speed into power output. More than 1 year of data ranging from March 2001 to May 2002 are available for the investigation.

7.4.1 Neuenkirchen

The wind farm Neuenkirchen has six wind turbines with a hub height of 70 m and a rated power of 1 MW, each located about 16 km south-east of the town of Cuxhaven.

The wind power predictions are calculated with and without the stability correction. First, the ratio between predicted and measured power output versus the inverse stability parameter z/L shows the effect of using the corrected wind profile compared with the neutral one. The predicted power considerably underestimates the measured power in stable situations if the neutral profile is used (Fig. 7.17 (left)). Therefore, the influence of thermal stratification on the power output is clearly visible and not covered by other sources of error. In contrast to the wind speed, there is no systematic overestimation in unstable conditions. The thermal correction eliminates the systematic deviation for unstable stratification (Fig. 7.17 (right)). The scatter is still profound but rather symmetric around the optimal value 1.

Stable stratification mainly occurs at nighttime when the ground cools down and the decreased turbulent momentum transport leads to a weaker coupling of the surface layer to the geostrophic driving force as discussed in Chap. 3. In these situations the predicted wind speed is typically too high if the neutral wind profile is used to transfer the wind speed given by the NWP model at 10 m to hub height, so that the power output is on average overestimated as well. In unstable situations which are commonly encountered at daytime the neutral profile tends to underestimate the wind speed and, hence, the mean predicted power output is generally too low. This characteristic property of the neutral profile leads to a diurnal cycle of the bias of the

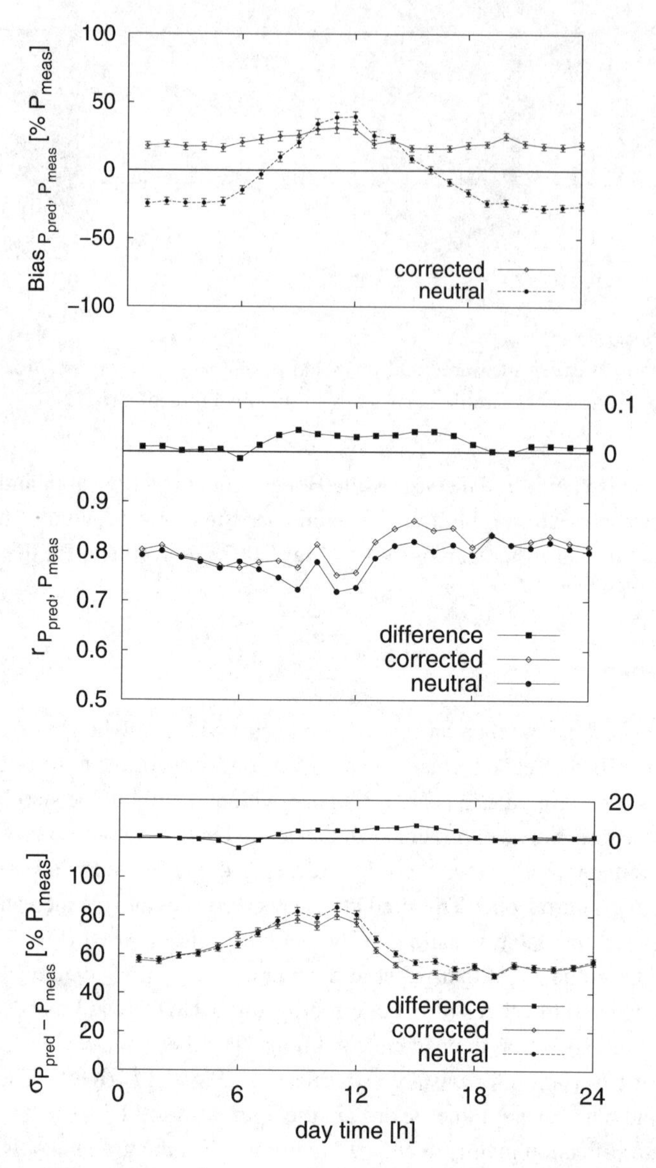

Fig. 7.18. Comparison of error of power prediction using the neutral profile and the thermally corrected profile for wind farm Neuenkirchen. As usual, the bias (*top*), cross-correlation (*middle*) and standard deviation of error (*bottom*) are shown. The upper part of each figure shows the difference between both methods

power prediction in Fig. 7.18 (top). During the night the relative bias is negative; it increases after sunrise with a maximum around noon and decreases again during the afternoon. If the thermal correction is applied the bias is increased at nighttime and decreased at daytime and almost completely levels out the diurnal variations.

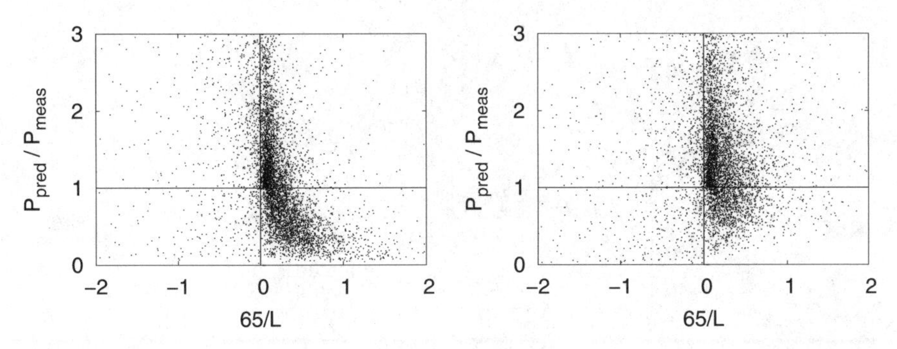

Fig. 7.19. Ratio between measured and predicted power output versus z/L for wind farm Hengsterholz using neutral profile (*left*) and with stability correction (*right*)

In addition, the average cross-correlation between predicted and measured power improves from 0.76 to 0.79. As Fig. 7.18 (middle) shows, this is mainly due to an increase during the day and seems to correspond to the fact that the LM prediction of the temperature differences and, thus, the stability parameter is significantly better in this period of time (see Fig. 7.16 (middle)). The same holds for the relative standard deviation of error, which slightly decreases from 67% to 64% (Fig. 7.18 (bottom)).

7.4.2 Hengsterholz

The wind farm Hengsterholz consists of six Vestas V47 with hub height of 65 m and an installed power of 660 kW each. The power prediction is carried out in the same way as for Neuenkirchen.

As before, considering thermal stratification also leads to a correction of the systematic deviations in stable situations (Fig. 7.19). The relative bias of the power forecast shown in Fig. 7.20 (top) has a far more pronounced diurnal cycle at this site compared with Neuenkirchen and varies about 90% of the measured power. Large variations in the bias over the day are typical for inland sites in flat terrain if the neutral profile is used. This is basically for two reasons. First, the neutral profile does not reflect thermal stratification in calculating the wind speed at hub height based on the predicted wind speed given by the NWP model at 10-m height. Second, the NWP model does not accurately predict the diurnal cycle of the wind speed at 10 m. This systematic error leads to a diurnal variation in the bias of the wind speed prediction which is quite considerable for inland sites, as described in Chap. 6. Hence, at this site the stability correction can only reduce the diurnal cycle of the bias that is due the extrapolation to hub height but cannot level it out completely as the influence of the initial NWP wind prediction is still prominent.

However, the benefits of applying the correction are still considerable. The cross-correlation averaged over the prediction times of 1 day increases from 0.74 to 0.78

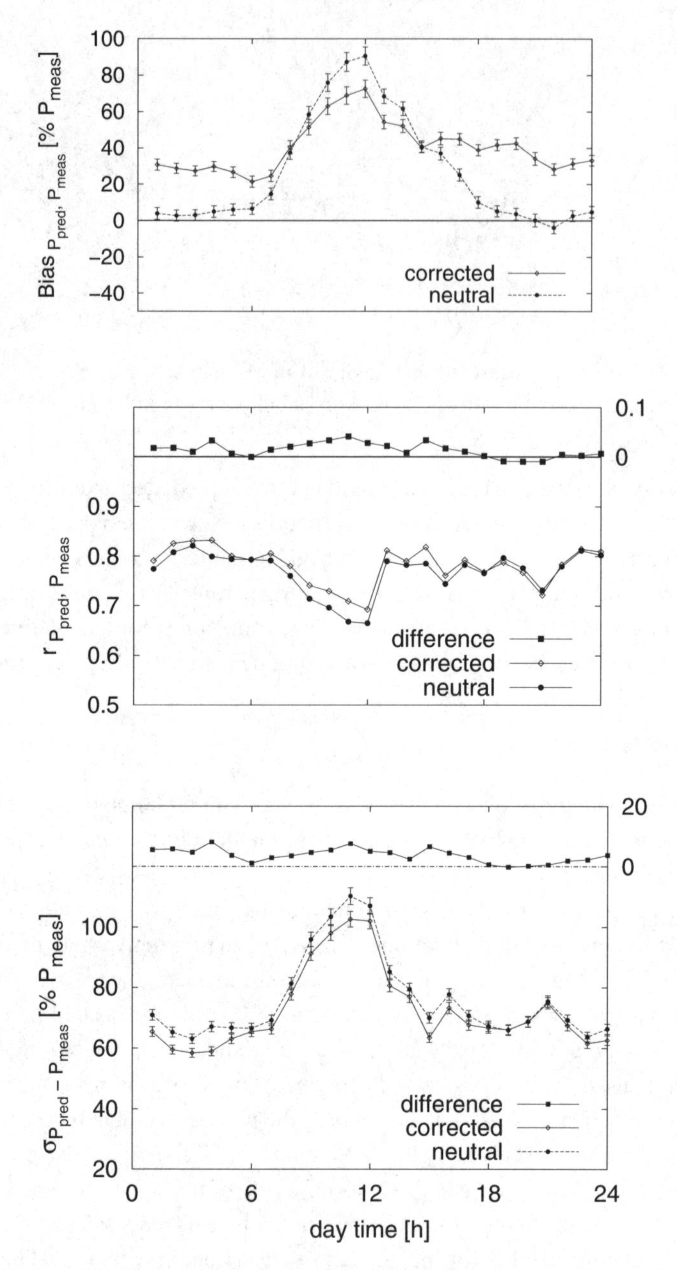

Fig. 7.20. Comparison of error of power prediction using the neutral profile and the thermally corrected profile for wind farm Hengsterholz. As usual the bias (*top*), cross-correlation (*middle*) and standard deviation of the error (*bottom*) are shown. The upper part of each figure shows the difference between both methods

(Fig. 7.20 (middle)). Moreover, the relative sde drops from 81% to 73% (Fig. 7.20 (bottom)). Compared to Neuenkirchen, improvements are not restricted to a particular time of day.

7.5 Conclusion

This chapter deals with an essential problem of wind power forecasting, namely transfering the wind speed at a given height, typically in the range 10–40 m, to the hub height of a wind turbine. The investigation shows that the thermal stratification of the atmosphere in the surface layer has to be considered. Applying stability corrections to the commonly used logarithmic wind profile leads to significant improvements in the calculation of the wind speed at hub height and, hence, can be benefitially implemented into physical power prediction systems.

The thermal corrections are based on the Monin–Obukhov theory, which proves to be successfully applicable not only in special stationary meteorological situations but also in general. The error which occurs by extrapolating the wind speed to hub height can be significantly reduced, in certain cases by 50%. For the investigated hub height of 80 m the error reduction does not depend on the height of the given wind speed. Of course, this result has to be confirmed for other sites with different local conditions. Moreover, the choice of the roughness length, which is subjective in many cases, does not have a large impact on the improvement achieved by the stability correction. The correction scheme is, as expected, limited to the surface layer comprising about the first 100 m of the atmosphere where the logarithmic profile is valid. At the investigated site a successful correction is possible for heights up to 80 m.

To determine the Monin–Obukhov length L, temperature differences between different heights are required. It is shown that taking into account the difference from the upper part of the surface layer between 20 m and 80 m leads to better results than estimating the temperature gradient based on low heights.

A crucial point regarding the implementation of the stability correction into a power prediction system is the predictability of the temperature differences by the NWP system Lokalmodell (LM) of the German Weather Service. Though the LM is able to predict the type of thermal stratification, i.e. whether the situation is stable or unstable, it does not accurately determine the degree of stability. In particular, extremely stable conditions are underestimated by too small temperature gradients, while extremely unstable situations are overestimated. Hence, the prediction of thermal stratification by the NWP model is not optimal and works relatively better for daytime forecasts than during the night.

Nevertheless, the thermal corrections of the wind profile, including the predicted temperature differences of LM are implemented into *Previento* and tested at two sites. The benefits of the correction are clearly visible in the power prediction despite other sources of error that might overcast the effect. The diurnal cycle of the bias can largely be reduced, particularly for sites with a weaker amplitude of the diurnal variations. Improvements in the forecast accuracy are observed for mainly the daytime hours, and this seems to be due to the fact that the LM prediction of thermal stability performes better in this period of time.

The strong reduction in the error in calculating the wind speed at hub height because of the consideration of thermal stratification leads to an overall improvement of about 5% in the error of the power prediction.

8

Assessment of Wind Speed Dependent Prediction Error

Abstract. The investigations in this chapter follow the idea that the prediction error quantitatively depends on the meteorological situation that has to be predicted. As a first approach the wind speed as a main indicator of the forecast situation is considered in greater detail. The probability density functions (pdf) of the measured wind speed conditioned on the predicted one are found to be Gaussian in the range of wind speeds that is relevant for wind energy applications. An analysis of the standard deviations of these conditional pdfs reveals no systematic dependence of the accuracy of the wind speed prediction on the magnitude of the wind speed. With the pdfs of the wind speed as basic elements, the strongly non-Gaussian distribution of the power prediction error is explained underlining the central role of the non linear power curve. Moreover, the power error distribution can easily be estimated based on the statistics of the wind speed, the wind speed forecast error and the power curve of the turbine. Thus, it can be reconstructed without knowing the actual measured power output, which is interesting for future sites or sites where no data are available. In addition, a simple formula based on linearising the standard deviation of error is derived. This model illustrates the dominating effect of small relative errors in the wind speed prediction being amplified by the local derivative of the non-linear power curve.

8.1 Idea Behind Detailed Error Assessment

The standard error measures that are used in the previous chapter are based on annual averages of the data and provide only one constant value for each forecast time. However, there is reason to believe that the magnitude of the error quantitatively depends on the meteorological situation to be predicted. Thus, a more detailed view on the prediction error is required, where the main parameters that characterise typical wind conditions have to be identified and related to the corresponding forecast error.

The first parameter that can serve as an indicator of the forecast situation is the wind speed itself. It is a continuous parameter closely related to the forecast situation and the main input into the power prediction system. In this chapter the role of the predicted wind speed and its prediction error are investigated more deeply.

As outlined in Chap. 4 the chain of events seems rather straightforward: the initial uncertainty introduced by the error of the wind speed prediction is propagated through the power prediction system where it is mainly subject to changes by the power curve. Due to its non-linearity the power curve is expected to amplify or damp initial errors in the wind speed according to its local derivative. And this derivative is a function of the wind speed. So, clearly, the power curve is expected to be the key element that connects the errors of the wind speed prediction and the power prediction. Thus, the major aim here is to derive a quantitative relation between the two.

The practical use of this relation and its implementation into a wind power prediction system is to provide the user with additional information to estimate the risk of trusting in the prediction. Hence, the prediction system has to supply the prediction itself and a useful indication concerning the reliability of the individual prediction. As financial losses might be proportional to the magnitude of the prediction error, the inherent risk of faulty predictions must be known for each forecast situation.

8.2 Introduction of Conditional Probability Density Functions

It was seen in Sect. 6.3 that the probability density function of wind speed error, $\mathrm{pdf}(\epsilon_u)$, is Gaussian in most cases. As mentioned above, a more detailed information concerning the error to be expected in special conditions is desirable. Thus, a first approach is to refine the pdfs and look at the statistical properties of the measured wind when the predicted wind speed is confined to a certain value which, from a mathematical point of view, leads to conditional pdfs. In this section the deviations of the measured wind speed from one is investigated in terms of these conditional pdfs.

Predicted and measured values of a meteorological variable at the same point of time and space are naturally not independent. In fact, they are supposed to be highly correlated as this is a major prerequisite for an accurate prediction. The pairs $(x_{\mathrm{pred}}, x_{\mathrm{meas}})$ are drawn simultaneously from a joint distribution, pdf$(x_{\mathrm{pred}}, x_{\mathrm{meas}})$, that characterises the statistical properties of the prediction and its error. This means that for arbitrary but fixed predicted values x_{pred}, the occurrences of the corresponding measured values x_{meas} are expected to be mainly concentrated in an interval around x_{pred} rather than being spread over the whole range of all possible values.

In the following investigation the “prediction perspective” is taken, which means that the predicted values are used as condition to the measurements. This aims at formulating the equations so that they can directly be used for prediction purposes.

Formally the conditional pdf of the wind speed is given by

$$\mathrm{pdf}(u_{\mathrm{meas}}|u_{\mathrm{pred}}) = \frac{\mathrm{pdf}(u_{\mathrm{pred}}, u_{\mathrm{meas}})}{\mathrm{pdf}(u_{\mathrm{pred}})} , \tag{8.1}$$

where $\mathrm{pdf}(u_{\mathrm{pred}}, u_{\mathrm{meas}})$ is the joint distribution and $\mathrm{pdf}(u_{\mathrm{pred}})$ is the unconditional, so-called marginal, wind speed distribution.

For practical purposes the time series are given by a finite number of data points with a certain accuracy. So the probability function of the measured wind speed u_{meas} under the condition u_{pred} is approximately calculated by confining the prediction values to an interval around u_{pred} and bin-counting the corresponding values for u_{meas}.

The conditional pdfs can, of course, be used to obtain the unconditional distribution by

$$\mathrm{pdf}(u_{\mathrm{meas}}) = \int_0^{\infty} \mathrm{pdf}(u_{\mathrm{meas}}|u_{\mathrm{pred}})\, \mathrm{pdf}(u_{\mathrm{pred}})\, du_{\mathrm{pred}} \,. \tag{8.2}$$

These conditional pdfs of the wind speed can be used to reconstruct the error distribution, pdf(ϵ_u), of the wind speed prediction that was found in Chap. 6. The conditional pdfs defined in (8.1) already contain all the information needed for this as they describe the deviations between predicted and measured values. With the transformation

$$u_{\mathrm{meas}} \mapsto \epsilon_u, \qquad \text{where} \quad \epsilon_u = u_{\mathrm{pred}} - u_{\mathrm{meas}} \,, \tag{8.3}$$

the pdf($u_{\mathrm{meas}}|u_{\mathrm{pred}}$) are merely mirrored and shifted along the abscissa and transferred to pdf($\epsilon_u|u_{\mathrm{pred}}$). The original, unconditional error distribution can then be recovered by

$$\mathrm{pdf}(\epsilon_u) = \int_0^{\infty} \mathrm{pdf}(\epsilon_u|u_{\mathrm{pred}})\, \mathrm{pdf}(u_{\mathrm{pred}})\, du_{\mathrm{pred}} \,. \tag{8.4}$$

Note that if the conditional pdfs are approximately equal to each other and do not vary much with u_{pred}, one obtains $\mathrm{pdf}(\epsilon_u) \approx \mathrm{pdf}(\epsilon_u|u_{\mathrm{pred}})$. In this case the overall pdfs of the error can directly serve as conditional pdf.

8.2.1 Reconstructing the Distribution of the Power Prediction Error

Knowing the statistical properties of the wind speed is very helpful with regard to the prediction of the power output as the wind speed is the main input variable fed into the power prediction system. However, as seen in Chap. 6, the error distributions of the power prediction differ qualitatively from those of the wind speed in that they are neither Gaussian nor approximately Gaussian, which is mainly due to the non-linear power curve of the wind turbines. The following investigation focusses on the role of the power curve in transforming the distributions of the wind speed and its error into those of the power output.

In Fig. 8.1 a typical power curve, $P(u)$, is shown with cut-in speed around 4 m/s, followed by a sharp increase of power over the interval 5–10 m/s and a saturation at the level of rated power for higher wind speeds. Let Δu be a small interval around

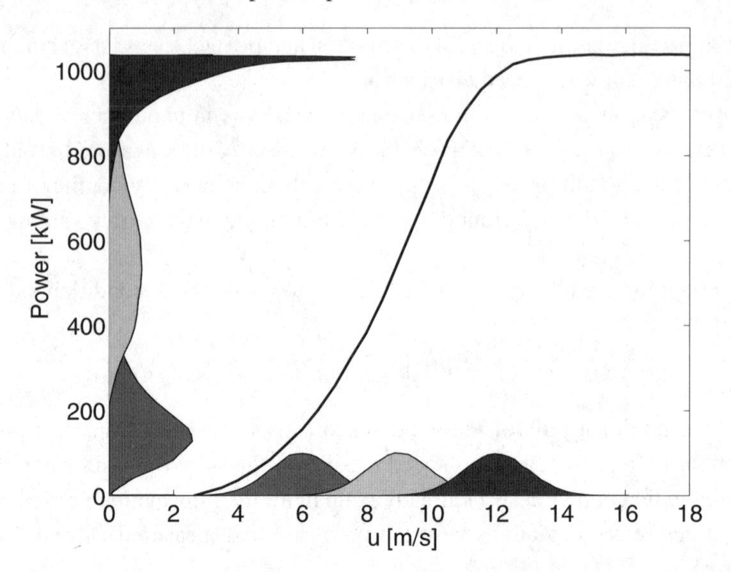

Fig. 8.1. Power curve of a pitch regulated wind turbine (*solid line*). For three different wind speeds, conditional pdfs, pdf($u_{\mathrm{meas}}|u_{\mathrm{pred},i}$), which have mean $= u_{\mathrm{pred,i}}$ and standard deviation 1, are illustrated on the x-axis. The corresponding pdf($P(u_{\mathrm{meas}})|P(u_{\mathrm{pred,i}})$) constructed from (8.5) are plotted on the y-axis. The pdfs are not normalised for better visualisation. For small and large wind speeds the Gaussian wind speed distributions are strongly deformed and no longer symmetric. For medium wind speeds the pdf of the power is significantly flatter and broader than the pdf of the wind speeds

the wind speed u and $\Delta P = P(u + \Delta u) - P(u)$ be the resulting difference in the power output. For small Δu the corresponding ΔP is then given by a Taylor expansion around u:

$$\Delta P = \frac{\mathrm{d}P}{\mathrm{du}}(u)\Delta u\ . \tag{8.5}$$

This equation generally describes how wind speed intervals are mapped to power intervals. If Δu is regarded as a small deviation between predicted and measured wind speed, (8.5) illustrates that the power curve scales errors in the wind speed according to its local derivative. Thus, whether deviations in the wind speed are amplified or damped depends on the magnitude of the wind speed.

Using (8.1) the probability to find a measurement value u in the interval $[u_{\mathrm{meas}}, u_{\mathrm{meas}} + \Delta u]$ with u_{pred} confined to $[u_{\mathrm{pred}}, u_{\mathrm{pred}} + \Delta u]$ is

$$w := \mathrm{pdf}(u_{\mathrm{meas}}|u_{\mathrm{pred}})\,\mathrm{pdf}(u_{\mathrm{pred}})\,(\Delta u)^2\ . \tag{8.6}$$

If the area $(\Delta u)^2$ around $(u_{\mathrm{meas}}, u_{\mathrm{pred}})$ is mapped to $(\Delta P)^2$ around $(P(u_{\mathrm{meas}}), P(u_{\mathrm{pred}}))$ according to (8.5), the probability w is preserved because all events that are recorded in the wind speed intervals also occur in the power output intervals. Hence,

$$
\begin{aligned}
w &= \mathrm{pdf}(u_{\mathrm{meas}}|u_{\mathrm{pred}})\,\mathrm{pdf}(u_{\mathrm{pred}})\,(\Delta u)^2 \\
&\overset{!}{=} \mathrm{pdf}(P(u_{\mathrm{meas}})|P(u_{\mathrm{pred}}))\,\mathrm{pdf}(P(u_{\mathrm{pred}}))\,(\Delta P)^2 \\
&= \mathrm{pdf}(P(u_{\mathrm{meas}})|P(u_{\mathrm{pred}}))\,\mathrm{pdf}(P(u_{\mathrm{pred}})) \\
&\quad \times \left(\frac{\mathrm{d}P}{\mathrm{du}}(u_{\mathrm{meas}})\Delta u\right)\left(\frac{\mathrm{d}P}{\mathrm{du}}(u_{\mathrm{pred}})\Delta u\right).
\end{aligned} \tag{8.7}
$$

If (8.7) is solved for the desired pdfs of the power output one obtains

$$
\begin{aligned}
&\mathrm{pdf}(P(u_{\mathrm{meas}})|P(u_{\mathrm{pred}}))\,\mathrm{pdf}(P(u_{\mathrm{pred}})) \\
&\quad = \mathrm{pdf}(u_{\mathrm{meas}}|u_{\mathrm{pred}})\,\mathrm{pdf}(u_{\mathrm{pred}})\left(\frac{\mathrm{d}P}{\mathrm{du}}(u_{\mathrm{meas}})\right)^{-1}\left(\frac{\mathrm{d}P}{\mathrm{du}}(u_{\mathrm{pred}})\right)^{-1}.
\end{aligned} \tag{8.8}
$$

This equation provides the essential relation between the distributions of wind speed on the one hand and power on the other. As expected, the power curve plays a crucial role in connecting the statistical properties of both quantities. Note that the divergence in (8.8) for $(\mathrm{d}P/\mathrm{d}u)^{-1} \to \infty$ is compensated by $\Delta P \to 0$ according to (8.5) so that the probability w is always well defined.

Figure 8.1 illustrates the effect of (8.5) for three different Gaussian wind speed distributions with same standard deviations. For small and large wind speeds the Gaussian wind speed distributions are strongly deformed and no longer symmetric. For medium wind speeds the pdf of the power is significantly flatter and broader than the pdf of the wind speeds.

The next steps towards a full description of the statistics of the power prediction error in terms of the wind speed distributions are now right ahead: Equation (8.8) is used to create the functions $\mathrm{pdf}(P(u_{\mathrm{meas}})|P(u_{\mathrm{pred}}))$ for each $P(u_{\mathrm{pred}})$. Then these functions are weighted according to the frequency distribution of $P(u_{\mathrm{pred}})$ and their variables are shifted analogous to transformation (8.3). Finally, these shifted functions are added and the unconditional pdf of the power prediction error is reconstructed.

Equations (8.7) and (8.8) suggest that the basic elements in this reconstruction procedure can be defined by

$$
\begin{aligned}
F(P(u_{\mathrm{meas}}), P(u_{\mathrm{pred}})) &:= \mathrm{pdf}(u_{\mathrm{meas}}|u_{\mathrm{pred}})\,\mathrm{pdf}(u_{\mathrm{pred}}) \\
&\quad \times \left(\frac{\mathrm{d}P}{\mathrm{du}}(u_{\mathrm{meas}})\right)^{-1}\left(\frac{\mathrm{d}P}{\mathrm{du}}(u_{\mathrm{pred}})\right)^{-1}.
\end{aligned} \tag{8.9}
$$

Note that this definition is an intermediate step that conveniently summarises all mathematical terms involved. The functions F contain the information how the conditional pdf of the wind speed has to be transformed to the corresponding power pdf and which weight this pdf has. Once the right-hand side of (8.9) has been used, F is defined in the power domain, i.e. the independent variables are $\tilde{P}_{\mathrm{meas}} := P(u_{\mathrm{meas}})$ and $P_{\mathrm{pred}} := P(u_{\mathrm{pred}})$.

The notation $\tilde{P}_{\text{meas}}$ denotes the measured power obtained by plugging the measured wind speed at hub height into the theoretical power curve. Generally, $\tilde{P}_{\text{meas}}$ slightly differs from the direct measurement of the power output, P_{meas}, because the power curve of the local wind turbine might deviate from the theoretical curve or additional errors that are not covered by the power curve come in.

Analogous to the transformation of the conditional wind speed distribution in (8.3), the first variable in each of the functions F is shifted according to

$$P_{\text{meas}} \mapsto \epsilon_P, \quad \text{where} \quad \epsilon_P = P_{\text{pred}} - \tilde{P}_{\text{meas}} \,. \tag{8.10}$$

In the final step the unconditional distribution of the power prediction error is obtained by the integration

$$\text{pdf}_{\text{rec}}(\epsilon_P) = \int_0^\infty F(\epsilon_P, P_{\text{pred}})\, dP_{\text{pred}} \,. \tag{8.11}$$

Despite having a slight touch of looking complicated, this method has some benefits. It exclusively provides the error statistics of the power prediction based on three ingredients: the conditional distributions of the wind speed prediction, the derivative of the power curve and the distribution of the predicted wind speeds. For practical purposes these three components are either given or can be estimated. The power curve is typically known numerically so that the derivative can easily be obtained. The distribution of the predicted wind speed is normally provided by the NWP output, but if not, it can be estimated from prediction data of nearby sites or from Weibull distributions of the measurement data. If the conditional pdfs of the wind speed are not known for the site in question, it can be assumed that they are Gaussian with mean values and standard deviations obtained by a qualified guess.

The concepts developed in this section are now applied to data from real sites to check if the relations between the wind speed and the power distributions can be confirmed with finite sets of data points.

8.3 Conditional Probability Density Functions of Wind Speed Data

The conditional pdfs of the wind speed are calculated as described by (8.1), i.e. the distribution of the measured values is determined with the corresponding predicted value confined to a certain interval. The range of the occurring u_{pred} is divided into equidistant bins with width Δu which is typically set to 1 m/s. The boundaries of the bins are given by $[u_{\text{pred,i}} - \Delta u/2, u_{\text{pred,i}} + \Delta u/2]$. Hence, $u_{\text{pred,i}}$ denotes the middle of the bins.

In Figs. 8.2 and 8.3 the conditional pdfs of the wind speed at two sites with different average wind speeds are shown based on measured and predicted data for the year 1996. In these examples the 36-h values of prediction and measurement are used, but the qualitative behaviour of the other prediction times is comparable.

The distributions for small wind speeds are unsymmetric and far from being normal. This is due to the fact that the wind speed is always positive and deviations from $u_{\mathrm{pred,i}}$ are limited towards lower wind speeds but not towards larger ones. With increasing $u_{\mathrm{pred,i}}$ the pdfs become rather symmetric.

Compared to a normal distribution with same mean and standard deviation, these pdfs can be considered as being approximately Gaussian although the number of data points available per pdf is relatively small. Again the two statistical tests from Chap. 6 are used to systematically check the normality of the conditional pdfs. As expected the hypothesis that the pdf is Gaussian is rejected by the χ^2 test and the Lilliefors test for distribution functions with small $u_{\mathrm{pred,i}}$. For larger prediction values both tests indicate normality: in Fig. 8.2 for the pdfs with $2.5\ \mathrm{m/s} \leq u_{\mathrm{pred,i}} \leq 6.5\ \mathrm{m/s}$, and in Fig. 8.3 for those with $2.5\ \mathrm{m/s} \leq u_{\mathrm{pred,i}} \leq 8.5\ \mathrm{m/s}$.

It is apparent from Figs. 8.2 and 8.3 that in most cases the mean of the conditional pdfs does not correspond to $u_{\mathrm{pred,i}}$, which is related to the systematic errors in the data that already occurred in Sect. 6.3. This is further illustrated in Fig. 8.4, where the means of pdf($u_{\mathrm{meas}}|u_{\mathrm{pred,i}}$) are plotted against $u_{\mathrm{pred,i}}$. For Rapshagen (Fig. 8.4 (top)) the mean values are mostly above the diagonal, i.e. the predictions are on average smaller than the measurements, leading to a negative bias (c.f. Fig. 6.7). In contrast to this the majority of mean values for Fehmarn (Fig. 8.4 (bottom)) are below the diagonal, and this corresponds to an overall positive bias.

For $u_{\mathrm{pred,i}} > 2\ \mathrm{m/s}$ the mean values increase linearly with $u_{\mathrm{pred,i}}$ but with slopes different from unity, indicating that the predictions systematically underestimate or overestimate the measurements. Within their error bars the mean values follow a line given by linear regression of all data points except for large $u_{\mathrm{pred,i}}$ where only few data points are available. Thus, the mean values of the conditional pdfs behave as expected, in that they reflect, on a bin-wise level, the systematic errors of the complete time series.

The pdfs in Figs. 8.2 and 8.3 seem to become wider for increasing $u_{\mathrm{pred,i}}$, suggesting that the deviations of u_{meas} from $u_{\mathrm{pred,i}}$, i.e. the prediction errors, grow on average; this is equivalent to a decreasing forecast accuracy. This directly leads to the question whether the accuracy of the wind speed prediction depends on the magnitude of the wind speed. In this investigation no final answer can be given as the behaviour for different sites and prediction times is rather inconsistent.

Consider the two examples for Rapshagen (Fig. 8.5 (top)) and Fehmarn (Fig. 8.5 (bottom)). While at Rapshagen (top) the relative standard deviation of pdf($u_{\mathrm{meas}}|$ $u_{\mathrm{pred,i}}$) has no apparent trend and oscillates around the corresponding relative sde (see Sect. 6.5), the pdf of Fehmarn (bottom) seems to have an increasing standard deviation. But such a clear trend cannot be detected for other prediction times at this site.

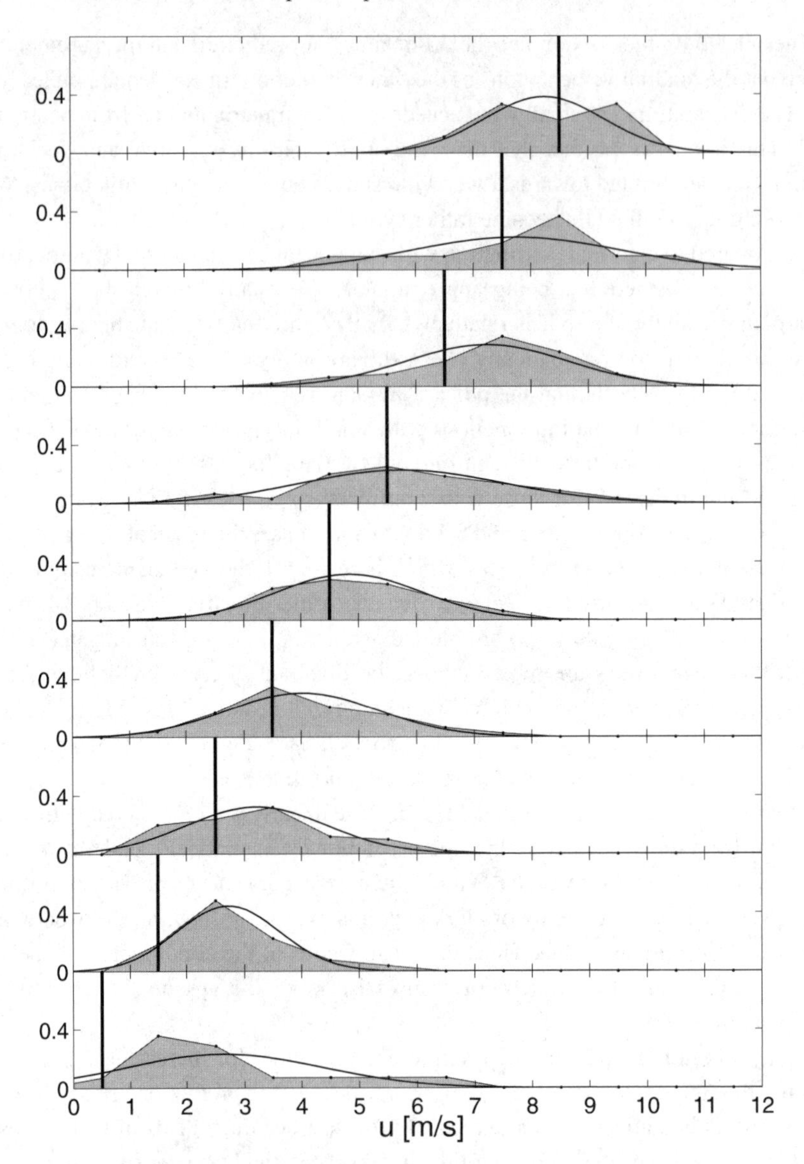

Fig. 8.2. Conditional pdfs of 36-h wind speed values for a site with low average wind speed (Rapshagen). The different pdf($u_{\mathrm{meas}}|u_{\mathrm{pred,i}}$) (*shaded areas*) are stacked, where $u_{\mathrm{pred,i}}$ has been varied in steps of 1 m/s, i.e. $\Delta u = 1$ m/s. The vertical line in each plot indicates the corresponding $u_{\mathrm{pred,i}}$. For comparison a normal distribution having the same mean and standard deviation as pdf($u_{\mathrm{meas}}|u_{\mathrm{pred,i}}$) is shown (*solid lines*). For pdfs with 2.5 m/s $\leq u_{\mathrm{pred,i}} \leq$ 6.5 m/s, the statistical tests indicate a Gaussian distribution. Note that conditional pdfs for large $u_{\mathrm{pred,i}}$ containing very few data points have been omitted

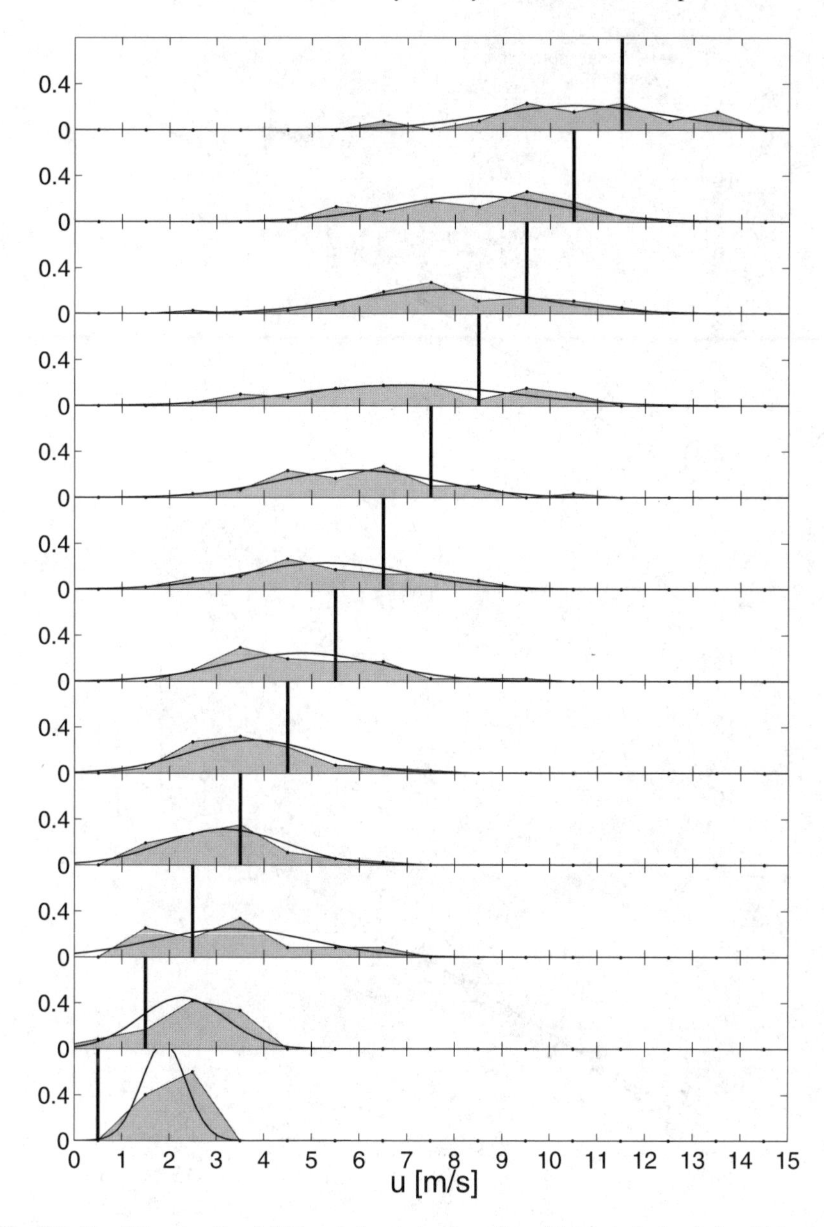

Fig. 8.3. Conditional pdfs of 36-h wind speed values for a high wind speed site (Fehmarn). Again the pdfs tend to become similar to normal distributions with increasing $u_{\mathrm{pred,i}}$, which is confirmed by the tests for pdfs with 2.5 m/s $\leq u_{\mathrm{pred,i}} \leq$ 8.5 m/s. Note that the number of data points decreases for larger $u_{\mathrm{pred,i}}$ and conditional pdfs containing very few data points have been omitted

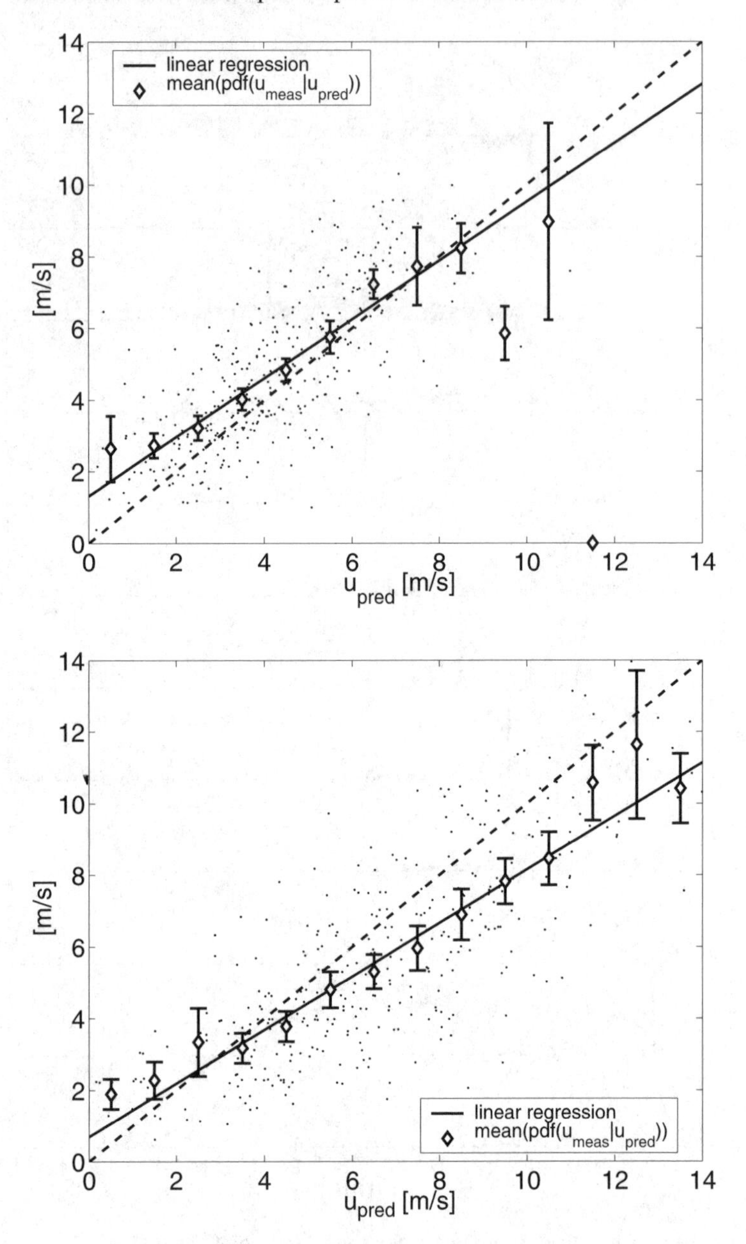

Fig. 8.4. The mean of pdf($u_{\text{meas}}|u_{\text{pred,i}}$) versus $u_{\text{pred,i}}$ together with the actual data points $(u_{\text{pred}}, u_{\text{meas}})$ for the 36-h prediction: *top*, Rapshagen (cf. Fig. 8.2); *bottom*, Fehmarn (cf. Fig. 8.3). In both cases the mean values are located on the linear regression line based on the data points for $u_{\text{pred}} > 2$ m/s. Their deviation from the diagonal (*dashed line*) reflects the systematic errors

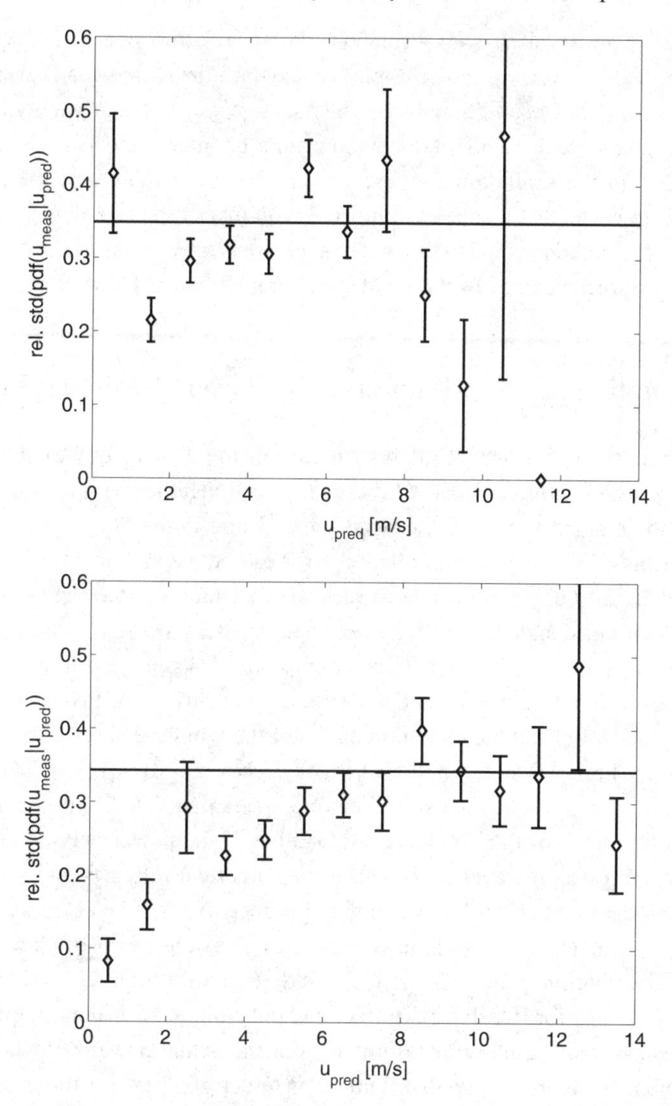

Fig. 8.5. The standard deviations of pdf($u_{\mathrm{meas}}|u_{\mathrm{pred,i}}$) normalised by the mean measured wind speed versus $u_{\mathrm{pred,i}}$ for Rapshagen (*top*) and Fehmarn (*bottom*). The prediction time is again 36 h. The solid line illustrates the unconditional relative standard deviation of error (cf. Fig. 6.8). The error bars provide the confidence intervals in calculating the standard deviation of the conditional pdfs. While for Rapshagen no clear trend is detectable there seems to be an increase of the relative standard deviations of the conditional pdfs with the predicted wind speed for Fehmarn. However, this is not systematic for neither this prediction time at other sites nor the other lead times at Fehmarn

Generally, most sites show some variation of the standard deviation of $\mathrm{pdf}(u_{\mathrm{meas}}|u_{\mathrm{pred,i}})$ over $u_{\mathrm{pred,i}}$ with relative standard deviations that deviate by the order of 0.1 from the unconditional sde. The pdfs for $u_{\mathrm{pred,i}} \leq 2$ m/s are typically unsymmetric, so that their standard deviation cannot be interpreted as 68% confidence interval. Due to the limitation of these pdfs at the lower boundary, their standard deviation is expected to be smaller compared with that of the symmetric pdfs.

These considerations lead to the result that there is only a weak, if any, systematic dependence of the accuracy of the wind speed prediction on the wind speed.

8.4 Estimating the Distribution of the Power Prediction Error

In this section it is shown that the distribution of the power output can indeed be derived from the conditional pdfs of the wind speed together with the power curve. Based on the empirical pdfs of the wind speed found in the Sect. 8.3 the unconditional distribution of the power prediction error can be reconstructed.

First of all, the quality of the reconstruction procedure is tested against a synthetic error distribution, denoted as $\mathrm{pdf}_{\mathrm{test}}(\epsilon_P)$, generated by using $\epsilon_P := P(u_{\mathrm{pred}}) - P(u_{\mathrm{meas}})$, i.e. using the theoretical power curve only. As Fig. 8.6 illustrates for the case of Rapshagen, the reconstruction of the error distribution that was calculated with (8.9) and (8.11) and the conditional pdfs of the wind speed obtained in the previous section (Figs. 8.2 and 8.3) almost exactly recovers $\mathrm{pdf}_{\mathrm{test}}(\epsilon_P)$). Thus, though the procedure for first decomposing the wind speed data into conditional distributions, scaling them with the reciprocal derivative of the power curve and reassembling everything again leaves some space for numerical artefacts and inaccuracies due to small data sets, it is robust enough to produce the expected results.

To be of practical use the more interesting test of the reconstruction is against the "real" distribution, $\mathrm{pdf}_{\mathrm{real}}(\epsilon_P)$, of the prediction error. Figure 8.7 shows the estimated $\mathrm{pdf}_{\mathrm{rec}}(\epsilon_P)$ for Rapshagen (top) compared with the distribution, $\mathrm{pdf}_{\mathrm{real}}(\epsilon_P)$, of the forecast error based on measurements of the actual power output. The overall agreement between the two distributions is rather good except for small ϵ_P. The reconstructed pdf covers the typical features of the original distribution in that it is unsymmetric in the same way and has the typical peak for small deviations.

As the error bars of $\mathrm{pdf}_{\mathrm{real}}(\epsilon_P)$ indicate, the reconstructed distribution does not perfectly match the real one in all of the bins, in particular for Fehmarn (Fig. 8.7 (bottom)). This deviation indicates additional sources of error that are not covered by considering the power curve effect only.

The results obtained so far explain how the power prediction errors are statistically distributed and why the distribution has this special shape. If measured and predicted data of the power output are available, there is no point in putting any effort in calculating the reconstructed $\mathrm{pdf}_{\mathrm{rec}}(\epsilon_P)$, as the error distribution is already

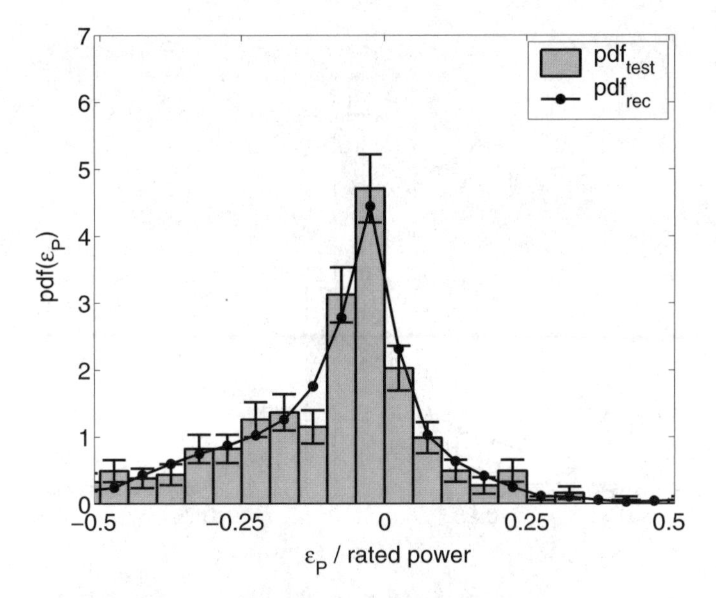

Fig. 8.6. Consistency check of the reconstruction procedure according to (8.9) and (8.11) for the 36-h prediction at Rapshagen. ϵ_P is normalised to the rated power. The reconstructed $\mathrm{pdf}_{\mathrm{rec}}(\epsilon_P)$ is compared to $\mathrm{pdf}_{\mathrm{test}}(\epsilon_P)$, which is based on evaluating $\epsilon_P := P(u_{\mathrm{pred}}) - P(u_{\mathrm{meas}})$. These two pdfs should be identical. However, they show small deviations for the bins around $\epsilon_P = 0$

available. However, if no data or only wind speed data are at hand, the reconstruction can be used to get an idea about how the error distribution of the power might look like. The minimum requirements to calculate this estimated distribution comprise four important aspects:

(1) the reasonable assumption that the conditional pdfs of the wind speed are all Gaussian (as discussed in Sect. 8.3) with the same standard deviation, $\sigma(\epsilon_u)$; second,
(2) a qualified guess concerning the value of $\sigma(\epsilon_u)$, e.g. from the weather service, as in [101], or nearby sites;
(3) the distribution of the wind speeds at the desired site; and
(4) availability of the power curve of the wind turbine.

With the considerations of this section the conditional pdfs, pdf($\epsilon_P|P_{\mathrm{pred}}$), can in principle be constructed and, hence, for each prediction value P_{pred} an individual estimate of the error distribution around this value could be supplied. However, the data sets that are used do not allow for a proper verification of the individual pdf($\epsilon_P|P_{\mathrm{pred}}$) with measured data because, in particular, for medium and high power outputs the number of data points is rather small. Hence, the difficulties in using

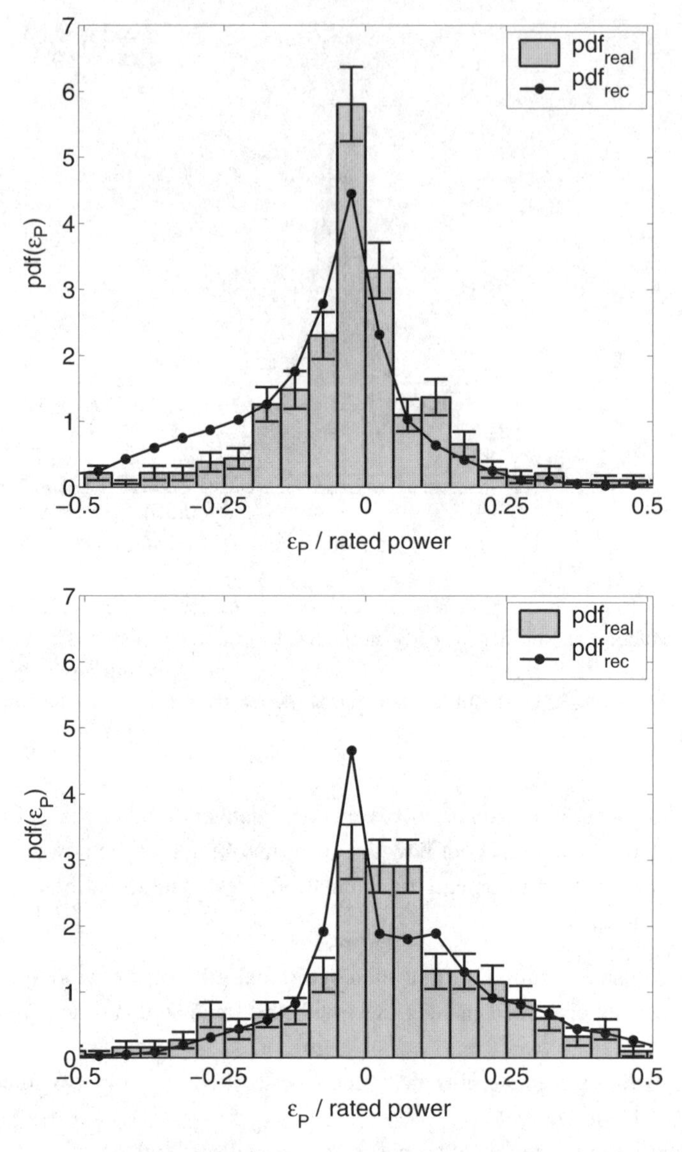

Fig. 8.7. Comparison of reconstructed $\mathrm{pdf}_{\mathrm{rec}}(\epsilon_P)$ according to (8.11) with the distribution of the actual measured power output $\mathrm{pdf}_{\mathrm{real}}(\epsilon_P)$ as it was recorded for the 36-h prediction at Rapshagen (*top*) and Fehmarn (*bottom*). ϵ_P is normalised to the rated power. As the error bars of the real distribution indicate, the agreement between the two pdfs for Rapshagen is rather good, with $\mathrm{pdf}_{\mathrm{rec}}$ covering the typical features of $\mathrm{pdf}_{\mathrm{real}}$ except the large peak for very small ϵ_P. The two pdfs at Fehmarn show differences for small positive ϵ_P, which indicate that the reconstruction model based on the power curve effect does not cover all error sources

statistical tests that already occurred for high wind speeds in Sect. 8.3 are more severe with regard to power. Thus, in this investigation, only the unconditional distribution of the power prediction error is reconstructed and compared to measurements because in this case all available data points at a site for a specific prediction time can be used.

In the next section it is shown that under simplifying assumptions a good estimation of the individual error bars for a specific wind power prediction can be derived.

8.5 Simple Modelling of the Power Prediction Error

Under the assumption that the prediction error of the underlying wind speed prediction does not change much over the range of typical wind speeds, as discussed in Sect. 8.3, the accuracy of the power forecast at a particular wind speed expressed as the standard deviation of error, sde, can be described rather well with a very simple approach.

For small deviations between predicted and measured wind speed, the sde of the power prediction is linearised using a Taylor expansion at the predicted wind speed u_{pred}, which leads to a "conditional" wind speed dependent error estimate given by

$$\sigma(\epsilon_P)|_{u_{\mathrm{pred}}} = \left|\frac{\mathrm{d}P}{\mathrm{d}u}\right|_{u_{\mathrm{pred}}} \overline{\sigma(\epsilon_u)} \,. \tag{8.12}$$

This equation is rather basic in the sense that it takes only the effect of the local derivative $\mathrm{d}P/\mathrm{d}u(u_{\mathrm{pred}})$ of the power curve into account and uses the average sde, $\overline{\sigma(\epsilon_u)}$, obtained from pdf($u_{\mathrm{meas}}|u_{\mathrm{pred,i}}$) as accuracy for all wind speeds. The approach neglects the fact that deviations between the certified, i.e. theoretical, power curve and the real power curve might occur which can have a significant influence on the prediction error. But this source of error can only be eliminated by using the corresponding measurement data to correct for these systematic deviations, while (8.12) generally models the error amplification effect of the power curve.

In Fig. 8.8 the error of the power prediction, where the corresponding predicted wind speed is confined to intervals of width 1 m/s for the two sites Rapshagen and Fehmarn, is plotted against the predicted wind speed. The bars denote the bin-wise standard deviation of the power prediction error, $\mathrm{sde}(\epsilon_P) = \sigma(\epsilon_P)|_{u_{\mathrm{pred}}}$, at a particular wind speed. Again, 1 year of WMEP data (see Chap. 5) have been used. Obviously, the accuracy of the power prediction depends on the predicted wind speed. This is mainly due to the power curve effect, as the solid line calculated from (8.12) indicates.

At Rapshagen this simple model describes rather precisely the behaviour of the actual power prediction error, as illustrated in Fig. 8.8 (top). However, at Fehmarn (Fig. 8.8 (bottom)) this modelling approach does not lead to an accurate description

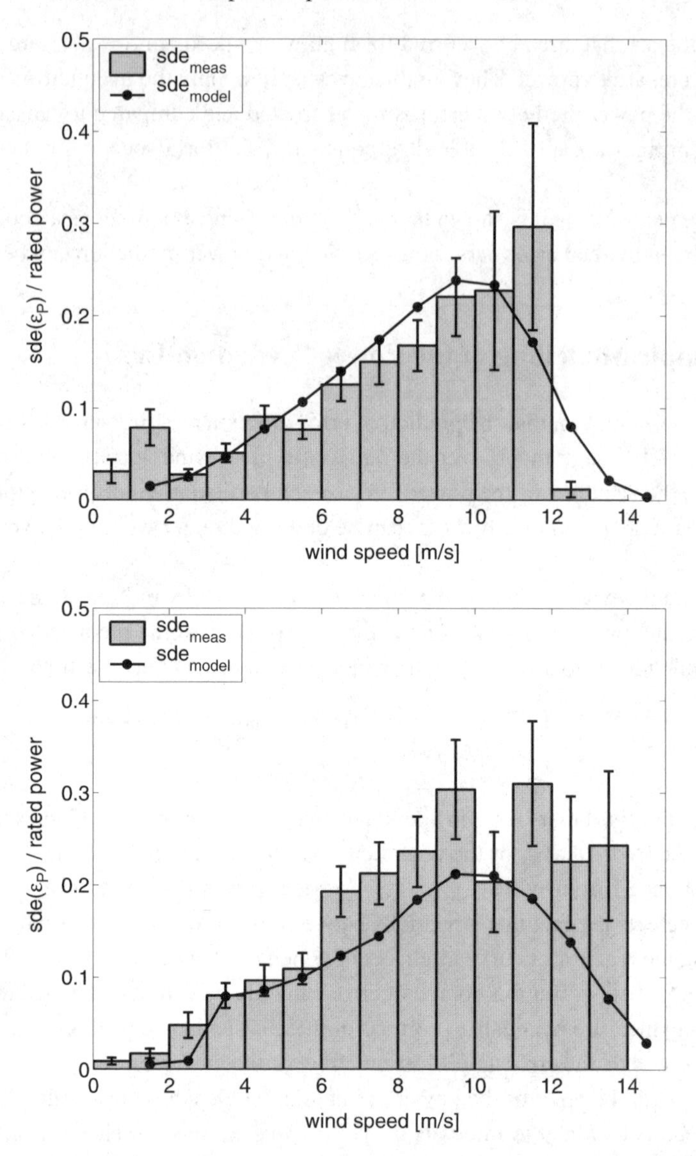

Fig. 8.8. Standard deviation of bin-wise power prediction error versus wind speed for the 36-h prediction at Rapshagen (*top*) and Fehmarn (*bottom*). The bars denote the sde between predicted and measured power output conditioned on wind speed intervals of width 1 m/s. For wind speeds up to 12 m/s the forecast error increases. At Rapshagen this behaviour is well approximated by the product of the derivative of the power curve and the wind speed error (*solid line*) according to (8.12), while at Fehmarn the simple modelling approach underestimates the power prediction error for larger wind speeds because of effects not covered by this model

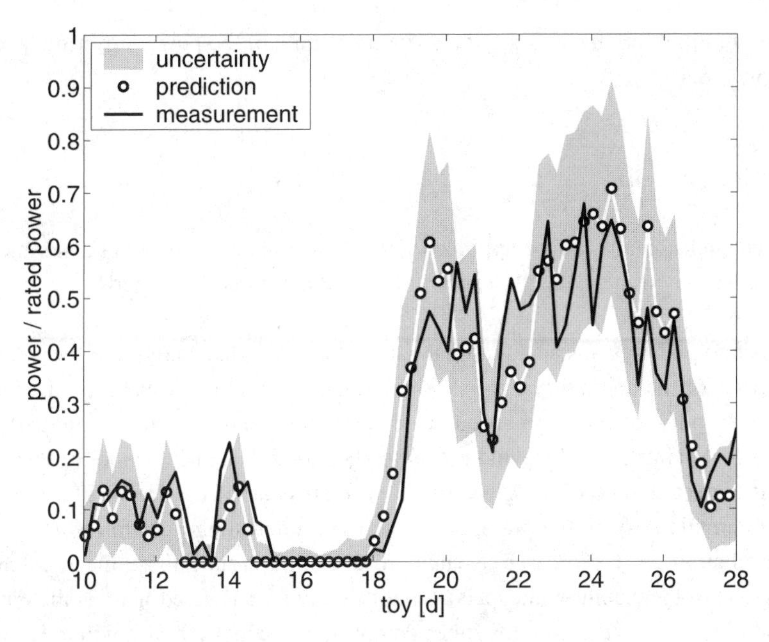

Fig. 8.9. Time series of prediction (○) and measurement (*solid line*) over a period of 18 days. Toy indicates the time of year in days. The *shaded* area is the uncertainty estimate given by (8.12). Typically, the uncertainty interval around the predicted value is small for low predictions, i.e. flat slope of the power curve, and large for predictions between 20% to 80% of the rated power where the power curve is steep. Of course, for power outputs near the rated power (not shown here) the uncertainty again decreases

of the prediction uncertainty. At this site the deviations between predicted and measured power output are not completely explained by the linear amplification of small wind speed errors according to (8.12) due to differences between the certified and the real power curve. But for many sites the model provides a rather good estimation of the wind speed dependent power prediction error.

The amplification of the error caused by the local slope of the power curve is clearly detectable in the sde of the power forecast and can be used to model the prediction error depending on the wind speed. Figure 8.9 demonstrates how this can be implemented in a power prediction system such as *Previento*. The specific uncertainty of each prediction given by (8.12) is illustrated by the shaded area around the predicted value. It is calculated by using the derivative of the certified power curve and the $\overline{\sigma(\epsilon_u)}$ corresponding to each prediction time. Thus, to use (8.12) for prediction purposes, historical data are needed to determine the statistical error, $\overline{\sigma(\epsilon_u)}$, of the underlying wind speed prediction.

This uncertainty interval provides additional information to users and enables them to assess the risk of a wrong prediction. Of course, this procedure has to be refined in the future by taking the unsymmetric pdfs of the power error into account (see Sect. 8.4).

8.6 Conclusion

This chapter contains a first approach towards a situation-dependent assessment of the prediction error where wind speed is the indicator that characterises the forecast situation.

Using wind speed as an additional parameter, a detailed analysis of the statistical properties of the prediction error shows that the conditional probability distribution functions (pdfs) of the wind speed error are mainly Gaussian in the range of wind speeds that is important for wind power applications. The mean values of these pdfs are, as expected, on the line given by linear regression of the scatter plot reflecting the systematic error in the prediction. The more interesting question as to whether the prediction error expressed as standard deviation around these mean values increases with increasing wind speed cannot finally be answered here as the data for the different sites and prediction times do not show a consistent behaviour. However, the results of this chapter suggest that there is only a weak, if any, systematic dependence of the prediction error measured as the standard deviation of the differences between prediction and measurement on the magnitude of the wind speed.

A main result of this chapter is that the non-Gaussian distribution of the power prediction error can be modelled. Understanding the mechanism that transforms initially Gaussian wind speed error distributions into strongly non-Gaussian power error distributions allows us to easily estimate the pdf of the power prediction error for any wind farm without knowing the actual predictions and measurements of the power output.

The approximated reconstruction of the pdf of the power forecast is based on three ingredients: the pdf of the measured wind speed conditioned on the predicted wind speed, the frequency distribution of the wind speed prediction and the power curve of the wind turbine. However, if for practical purposes the conditional pdf of the wind speed at a site is not available, it can be estimated by assuming that the distribution is Gaussian with a standard deviation of the wind speed error taken from a nearby site or from the publications of the local weather service (e.g. [101]). The exact frequency distribution of the predicted wind speed could be replaced by either the distribution of the measurements or distributions from nearby synoptic stations. This means that an estimate of the statistical distribution of the power prediction error at a site can be obtained using information that is readily available for existing wind farms and can even be obtained prior to the erection of the wind farm at the desired location. This is, for example, convenient to assess the prediction error for future offshore wind farms.

Of course, the "real" distributions of the power prediction errors given by comparing predicted and measured power output might look different. The method derived here does only consider the effect of the power curve, neglecting other sources of error. In particular, the power output is modelled by plugging wind speeds into the certified power curve representing the average behaviour of the wind turbine under standard conditions. However, experience shows that the output of individual turbines can well deviate significantly from their certified power curves for various reasons [77]. Thus, for practical use the limitation of this reconstruction is, clearly, that it is oblivious to these additional error sources beyond wind speed. Nevertheless, it sheds some light on the mechanism that produces the typical non-Gaussian distributions of the power prediction error.

A relatively simple model that describes the dependence of the power prediction error on the wind speed with a linearised approach is used. It shows that most of the error can be explained by the influence of the power curve that amplifies the rather constant prediction error of the wind speed according to its local derivative. This procedure can directly be implemented into previento to provide situation-dependent uncertainty estimates for each prediction time.

These uncertainty intervals are symmetric, which is a major shortcoming of this simple model because it is already clear that the distributions of the power prediction error are non-Gaussian and unsymmetric. However, to provide good estimations of the conditional power distributions around each prediction value, a lot of information concerning the statistics of the specific site has to be recorded. Therefore, the simple model suggested here is only the first step towards a comprehensive description of the situation-based error statistics.

9

Relating the Forecast Error to Meteorological Situations

Abstract. The investigation in this chapter focusses on the quantitative relation between the error of the wind speed prediction and the corresponding specific meteorological situation. With methods from synoptic climatology an automatic classification scheme is established using measurements of wind speed, wind direction and pressure at mean sea level to characterise the local weather conditions at a site. The classification procedure involves *principal component analysis* to efficiently reduce the data to the most relevant modes. *Cluster analysis* is used to group days with similar meteorological conditions into common classes. A comparison of these clusters with large-scale weather maps shows that typical weather situations are successfully captured by the classification scheme. The mean forecast error of the wind speed prediction of the German Weather Service is calculated for each of the clusters. It is found that different meteorological situations have indeed significant differences in the prediction error measured by a daily rmse, where the maximum rmse can be larger than the minimum rmse by a factor of 1.5 to 1.7 . As expected, high uncertainties in the forecast are found in situations where rather dynamic low-pressure systems with fronts cross the area of interest, while stationary high-pressure situations have significantly smaller prediction errors.

9.1 Introduction

The investigation in this chapter continues to follow the idea of evaluating the forecast error for specific weather situations. But in contrast to the previous chapter, the meteorological conditions will now be described by a far larger set of variables than only wind speed to include more details of the atmospheric state and its temporal evolution over 1 day. The aim is to really distinguish different weather classes and relate them quantitatively to their typical prediction errors.

It is a well-known fact that the performance of numerical weather prediction (NWP) systems is not equally well for every meteorological situation and that their accuracy depends on the situation that is to be forecast. In an overview of the prediction uncertainty of weather and climate forecasts, Palmer [87] points out that "certain types of atmospheric flow are known to be rather stable and hence predictable, others to be unstable and unpredictable". Thus, the challenge is to know in advance how predictable the current meteorological situation is.

There are already different approaches on how to include information about the changing reliability of the numerical forecast into the prediction. A very popular one is the use of ensemble predictions, where the chaotic properties of the non-linear equations of motion of the meteorological variables are exploited, e.g. described in [46, 87]. Lorenz [66] demonstrated that low dimensional non-linear weather models are sensitive to small changes in the initial conditions, and this is the typical indication of deterministic chaos. Hence, to get an overview of the possible range of weather situations that can evolve from a given situation, the initial condition of the NWP system is perturbed. Then the NWP system calculates separate predictions for each initial condition, leading to an ensemble of possible outcomes. Thus, ensemble forecasts provide a range of possible weather situations that can occur with a certain probability. The difficult part is to generate suitable ensembles allowing for a statistical interpretation of the results which might differ profoundly.

Ensemble forecasts are computationally expensive and, therefore, for operational use mainly generated by large meteorological institutes such as the European Centre for Medium Range Weather Forecasts (ECMWF). However, Landberg et al. [60] used a simpler version of ensemble predictions in connection with wind power application, which they denoted as "poor man's ensemble forecast". The idea is to take the spread between different prediction runs, e.g. at 00 UTC and 12 UTC, calculated for the same points of time in the future. The larger the deviation between the prediction runs, the greater the uncertainty of the prediction. This concept has recently been enhanced by Pinson and Kariniotakis [91, 92]. They established a continuous risk index from the spread of several forecast runs of the wind speed prediction to derive the corresponding degree of uncertainty connected to the power prediction.

The technique used in this chapter to assess the uncertainty of a specific prediction is different from the ensemble approach, as it is based on a classification scheme of the weather situation which is not directly related to the prediction system. The approach here is to describe the situation of each day by a suitable set of local meteorological variables, then sort it into a certain category and associate each of the categories with a typical prediction error derived from historical data.

Intuitively, at least two types of meteorological situations over Northern Europe are expected to show considerable differences in terms of the accuracy of the wind speed prediction. Weather situations with strong low-pressure systems coming in from the Atlantic Ocean are supposed to be of the unstable type that is hard to predict. These situations can be very dynamic as the advection speed of the low-pressure systems is typically rather large and, in addition, its frontal zones may cause strong winds. In dynamic cases the real situation can evolve quite differently from the one that had been predicted. In contrast to this, high-pressure systems with typically moderate wind speeds are rather stationary and can persist for several days, which should make it easier these to provide a reliable prediction.

It is the purpose of this chapter to establish a method that automatically generates useful weather classes and, hence, implicitly defines what "strong low" or "stationary high" means in terms of the local conditions at different sites. Moreover, the investigation here aims to answer the question how large the various prediction errors are and whether they differ significantly for different meteorological conditions.

9.2 Methods from Synoptic Climatology

The techniques that are applied in this investigation are well established in synoptic climatology. In a concise overview of the subject, Yarnal [124] points out that "synoptic climatology relates the atmospheric circulation to the surface environment". Hence, this branch of climatology explicitly aims at linking meteorological conditions to external variables, such as concentration of air pollutants, which is exactly the type of problem that is addressed in this work.

Quite a variety of powerful methods have been developed in this field and tested in numerous applications [5, 103, 124]. The main principles are very similar. The first step is to identify typical structures or patterns of the atmospheric circulation in order to develop a classification scheme that is only based on meteorological variables. After the classification is established the statistical or deterministic relationship between these structures and a surface variable of interest is investigated. Normally, this variable is non-meteorological in the sense that it does not describe the state of the atmospheric flow.

As regards the classification scheme, there are two main approaches: on the one hand, manual methods where a meteorologist evaluates the synoptic situation within a certain framework according to his or her experience; and on the other hand, automatic schemes which use computer-based algorithms to sort meteorological data into different classes. A well-known manual classification scheme for Central Europe is the catalogue of "Grosswetterlagen" by Hess and Brezsowski [34]. It comprises 29 large-scale weather situations which are distinguished by their different spatial arrangement of the pressure systems. One disadvantage in using this catalogue is that the large number of pre-defined weather classes requires times series with many data points in order to obtain statistically relevant results [63].

In contrast to this, automatic schemes typically exploit correlations between patterns or use eigenvector techniques such as principal component analysis (PCA) to extract synoptic weather classes from numerical data. Yarnal [124] points out that both the manual and the automatic methods contain a certain degree of human subjectivity because someone has to decide, e.g., on the set of variables and the number of classes.

The method chosen in this work to analyse how the forecast error is related to the weather type relies on computer-based classification techniques and follows in principle an investigation carried out by Shahgedanova et al. [103]. They used PCA in combination with cluster analysis of surface and upper air meteorological data to derive a classification scheme of the synoptic situation in Moscow; PCA was used for data reduction and cluster analysis to sort similar days into common groups. These weather types were then connected to typical concentrations of air pollutants such as CO and NO_2. It was found that certain weather patterns result in significantly high levels of urban air pollution.

In contrast to the method of Shahgedanova et al., the set of meteorological variables used here will be smaller with no upper air data involved. The role of the pollutant will be played by the forecast error. In the following, the use of PCA and cluster analysis is explained.

9.2.1 Principal Component Analysis

Principal component analysis (PCA) is used to extract the relevant eigenmodes from the meteorological data. This technique produces eigenmodes which can be ordered according to the degree of variance they explain in the data. Hence, the modes that contribute most to the time series can be selected for the further analysis which allows to effectively reduce the amount of data. The weather situation of each day can then be approximately expressed as a linear combination of these eigenmodes.

The variables used here are the horizontal wind vector $\vec{u} = (u, v)$ and the atmospheric pressure at mean sea level (pmsl). These are the natural candidates to start with as they are closely related to the wind field. To account for temporal variations over 1 day the variables are taken at 0, 6, 12, 18, 24 UTC. Several values of the wind vector per day resolve changes in wind direction and speed, e.g. during the passage of a frontal system. This provides the possibility of separating dynamic from static weather situations. The record of the temporal variations of pmsl can partly compensate for the fact that spatial pressure gradients which are the main driving force of the wind are not available in this investigation.

The temperature that is often included to find synoptic indices is not considered here. This is done because absolute temperature is mainly used to determine the type of air mass, e.g. by Shahgedanova et al. [103], but it is not directly related to the wind field. Moreover, temperature tends to be a dominating variable that requires the data sets to be split into winter and summer part. But doing so would leave only half the data points for the analysis, which makes it difficult to have significant results in the end. However, for further investigations it would be desirable to include temperature and humidity to be able to detect fronts with more parameters than changes in the wind vector. Of particular interest for further investigations are, of course, temperature differences at different heights to assess atmospheric stratification (c.f. [27]).

The measured data are written into a matrix M where the columns contain the different variables and each row corresponds to 1 day. The measurements at the various day times are considered as separate variables. As wind vector and surface pressure have very different orders of magnitude, both are normalised with their standard deviation and pmsl is additionally centralised by subtracting the mean value. Hence, M is given by:

$$M = \begin{pmatrix} u_{1,0} & \cdots & u_{1,24} & v_{1,0} & \cdots & v_{1,24} & \text{pmsl}_{1,0} & \cdots & \text{pmsl}_{1,24} \\ \vdots & & & & & & & & \vdots \\ u_{365,0} & \cdots & u_{365,24} & v_{365,0} & \cdots & v_{365,24} & \text{pmsl}_{365,0} & \cdots & \text{pmsl}_{365,24} \end{pmatrix}, \quad (9.1)$$

where the subscripts denote the day of the year and the time of day. M is the so-called data matrix with dimension 365×15 having one row for each day of the year with 15 measurements per day for the three variables at times 0, 6, 12, 18 and 24 UTC.

The PCA of matrix M is carried out numerically by diagonalising the so-called covariance matrix $C = M^t M$. The 15 eigenvectors, $\vec{p}_i$, of C are the desired principal components (PC) of M, while the eigenvalues, λ_i, of C are the weights of the different components that express the degree of variance a particular PC contains. A full description of this procedure can be found in [13].

The set of principal components $\{\vec{p}_i\}$ with $i = 1, \ldots, 15$ constitutes a new orthonormal basis of the phase space of the full set of data points. The ordering of the PC according to the eigenvalues, λ_i, allows to select only the first few PC spanning the subspace where most of the relevant dynamics takes place. The number of relevant modes, N, to be considered for further analysis is not precisely defined and has to be inferred from the spectrum of eigenvalues and the corresponding PC. In this investigation the visual inspection of these quantities and the cumulative variance $\sum \lambda_i$ of the first N eigenvalues will be used as criteria to decide on the number of modes to be used. Hence, by transforming to new coordinates and reducing the degrees of freedom, PCA efficiently codes the information contained in the meteorological raw data and in that sense acts as a data reduction technique.

The use of PCA as a data reduction technique means that the eigenvalue spectrum is mainly used to account for the degree of variance contained in each of the corresponding eigenmodes. This is a statistical or climatological interpretation of the eigenspectrum and not a non-linear systems approach. In the context of dynamic systems a similar technique, time-delay embedding, is also applied as a tool to decompose phase space dynamics in different modes with the aim of extracting the degrees of freedom of the non-linear system, e.g. described in detail by Broomhead and King [13] or Kantz and Schreiber [52]. However, the approach used in this work and in particular the construction of the data matrix M does not aim at providing a delay embedding of the time series as the sampling intervals are not adapted for this purpose and the number of data points is far too low. Hence, the number of eigenmodes provided in this context cannot contribute to the question how many degrees of freedom the weather or climate system has, as discussed, e.g., by Grassberger [37] and Nicholis and Nicholis [80].

After a choice concerning the number of relevant PC is made, the data in M is transformed to the reduced basis, so that each day can be represented as a linear combination of the selected PC. Let $\{\vec{q}_i\}$, $i = 1, \dots, N$, be the basis chosen from the first N PC of the full eigenvector basis, $\{\vec{p}_i\}$, of M. Then

$$Q := (\vec{q}_1 \dots \vec{q}_N) \tag{9.2}$$

is the transformation matrix that can be used to easily project the data in M on the new basis by a multiplication from the right:

$$X := MQ\,. \tag{9.3}$$

The entries in X are the scalar products $x_{ij} = \vec{m}_i \cdot \vec{q}_j$, where $\vec{m}_i$ is the ith row of M. In other words x_{ij} is the contribution of the jth PC to the ith day. Consequently, each day $\vec{m}_i$ can be approximately (because only an N-dimensional subspace is considered) expressed by

$$\vec{m}_i \approx \sum_{j=1}^{N} x_{ij}\vec{q}_j\,. \tag{9.4}$$

Thus, the 365 by N matrix X is the reduced data matrix that is thought to contain most of the relevant meteorological information of 1 year of data. However, so far nothing really happened in terms of the classification scheme because the data have merely been recoded. The next step applied to the reduced data set will be cluster analysis.

9.2.2 Cluster Analysis

Cluster analysis is a standard method used to group objects with similar properties together. As described in a concise overview by Everitt [23], it has a wide range of techniques and is often applied in climatological investigations [51, 103, 124].

In this work the aim is to obtain a rather small number of clusters which contain days with similar meteorological conditions being different from the days in the other clusters. Of course, the clusters found are required to represent typical weather classes.

The type of cluster analysis used here is called hierarchical or agglomerative. It acts iteratively on the phase space by computing the distances between each pair of objects in the phase space and then joining the closest two objects to a new cluster. After the new cluster is formed the procedure is repeated. Starting point is a situation where all points in the phase space are considered as separate clusters. As the final iteration merges all clusters into one group, the process has to be stopped when a certain number of clusters is reached. The criterion to stop the iterations is derived from observing the growing distances between the clusters.

Though the basic concept is rather straightforward there are profound differences in the results of various clustering procedures because the key point of this technique is how distances between clusters are defined and under which rule new clusters are formed. First of all, a metric has to be chosen that evaluates distances between single points in the phase space. Here the Euclidean metric (9.5) is applied as there is no apparent reason for a different one. A more crucial point is the definition of distances between clusters. As the type of distance measured implies which two clusters will be joined next, it is referred to as "linkage method". Three typical linkage methods are briefly introduced using the following notations.

Let $\vec{x}_i := (x_{i,1}, \ldots, x_{i,N})$, with $i = 1, \ldots, 365$, denote the coordinate vector describing a point in the N-dimensional reduced phase space. The distance $d(\vec{x}_i, \vec{x}_j)$ between two states is given by the Euclidean measure

$$d(\vec{x}_i, \vec{x}_j) = \sqrt{\frac{1}{N} \sum_{q=1}^{N} (x_{i,q} - x_{j,q})^2} \, . \tag{9.5}$$

After a number of N_A phase states has been joined into one cluster, C_A, the members of this cluster will be denoted as $\vec{x}_r^A$, where $r = 1, \ldots, N_A$ is the new index within C_A.

Complete Linkage Method

To evaluate the distance between two clusters C_A and C_B in the complete linkage method, all pairs of Euclidian distances between the members of C_A and C_B are computed. Then the maximum distance that is found between the individual members is used as the distance between the two clusters. Hence,

$$d_{\text{complete}}(C_A, C_B) := \max\left(d(\vec{x}_r^A, \vec{x}_p^B)\right), \quad \text{where} \quad (\vec{x}_r^A, \vec{x}_p^B) \in C_A \times C_B \, . \tag{9.6}$$

Average Linkage Method

Average linkage is a combination of complete linkage as described above and its complementary definition, called single linkage, where the minimum of the distance between cluster members is taken. In average linkage the mean value of the individual distances is used to define the cluster distance between two clusters C_A and C_B, i.e.

$$d_{\text{average}}(C_A, C_B) := \frac{1}{N_A N_B} \sum_{r=1}^{N_A} \sum_{p=1}^{N_B} d(\vec{x}_r^A, \vec{x}_p^B) \, , \tag{9.7}$$

where N_A and N_B are the respective numbers of elements in C_A and C_B.

Ward's Linkage Method

Ward's method is also referred to as minimum variance method [51] as it evaluates the change of within-cluster variance if two clusters are merged.

$$d^2_{\text{ward}}(C_A, C_B) := \frac{N_A \, N_B}{N_A + N_B} \; d\left(\overline{\vec{x}^A}, \overline{\vec{x}^B}\right)^2 , \tag{9.8}$$

where $\overline{\vec{x}^A} = \sum_{r=1}^{N_A} \vec{x}_r^A$ is the centre of mass of cluster C_A and $\overline{\vec{x}^B}$ is that of C_B.

Kalkstein et al. [51] thoroughly investigated average linkage, Ward's linkage and a third technique, centroid linkage, which is not used here. They found that for weather classification purposes, average linkage produced the "most realistic synoptic groupings" as it provides rather homogeneous groups of days with similar meteorological conditions and sorts extreme events into separate clusters instead of combining them into a common class. In contrast to the other techniques, average linkage minimises the within-cluster variance, i.e. the mean variance among the days within one cluster, and maximises the between-cluster variance, i.e. the mean variance between the centres of mass of different clusters. Shahgedanova et al. [103] also confirmed the usefulness of average linkage in their investigation.

Ward's linkage on the other hand tends to produce clusters of equal size and, therefore, sorts days with extreme weather conditions together with less extreme days, which "blurs distinctions between the types" [51]. Thus, this method was considered inferior compared with average linkage to produce meaningful synoptic classes. However, Yarnal [124] provides classification problems which are quite comparable to those provided by Kalkstein et al. [51] and Shahgedanova et al. [103] but where Ward's method is superior to average linkage, which leads to the conclusion that both methods should be tested.

In this work, average and Ward's linkage are used as they have been successfully applied in previous investigations. Additionally, complete linkage is also applied.

9.2.3 Daily Forecast Error of Wind Speed

For the analysis of the forecast error a new error measure is introduced that evaluates the performance of the prediction system over 1 day:

$$\begin{aligned} \text{rmse}_{\text{day}} &= \sqrt{\frac{1}{4} \sum_{t_{\text{pred}}} \left(u_{\text{pred},t_{\text{pred}}} - u_{\text{meas},t_{\text{pred}}}\right)^2}, \quad \text{with} \\ t_{\text{pred}} &= 6\,\text{h}, 12\,\text{h}, 18\,\text{h}, 24\,\text{h} \end{aligned} \tag{9.9}$$

where $u_{\text{pred},t}$ and $u_{\text{meas},t}$ are predicted and measured wind speed.

It is, of course, desirable to also include predictions with horizons beyond 24 h into the error measure. However, higher lead times have not been included so far because only two larger prediction times, 36 h and 48 h, are available in the investigated data set which is only half the number of data points for the analysis compared with the period 6 to 24 h. Hence, this first approach is restricted to the earlier lead times.

This "daily error" assigns one single value to each day in a cluster which is analogous to using the daily concentration of a pollutant by Shahgedanova et al. [103]. The characteristic property of this definition is that the point of time at which a deviation between prediction and measurement occurs is not relevant. Moreover, with the definition in (9.9) errors caused by coherent structures such as a wrongly predicted front are summarised in one error value, i.e. the correlation between succeeding deviations is implicitly taken into account.

The idea behind choosing $\mathrm{rmse}_{\mathrm{day}}$ can be illustrated by an example. Imagine that a low-pressure system approaches the domain of interest. It brings a rise in wind speeds that is predicted for the late afternoon, say 18 h, but actually arrives a few hours earlier, say 12 h. In this case the difference between predicted and measured values is rather large and negative at 12 h (because the prediction did not foresee the low) and small and negative at 18 h (because the prediction expected the wind speed to start increasing). Moreover, at 24 h the low has passed and the wind speed actually decreases while the prediction still suggests high wind speeds leading to a positive deviation. Now these situations can typically produce phase errors where the deviations at a number of lead times in a row are coherently affected. Another important point is that in terms of the daily error measure, it should not matter whether the low is too fast, as in this example, or is behind schedule, i.e. arrives later than predicted.

9.2.4 Tests of Statistical Significance

Resolving individual situations reduces the amount of data points available per cluster by a factor of about 10 and, thus, statistical significance becomes an important issue. Therefore, care has to be taken not to be fooled by statistical artefacts, so that in this chapter and the next, tests for statistical relevance will again be applied. To check if the differences in the statistical values of different clusters are not obtained by chance an F test together with Scheffe's multiple comparisons is applied which is based on relating the within-cluster variance of the error values to the variance of the error values in the remaining clusters. The details of the tests are described in Appendix B.

9.3 Results

The methods described above are applied to the data described in Chap. 5. To develop a climatological classification scheme, measured data from 1 year are used.

The horizontal wind vector $\vec{u}$ at a site is calculated by sine–cosine transformation of the measured wind speed and direction at, preferably, 30 m to avoid artefacts from obstacles that might occur for lower heights. However, if the 30-m measurement is not available, 10-m data are also used. The corresponding surface pressure is taken from the nearest synoptic station of the German Weather Service. It is converted to mean sea level using the barometric height formula.

To avoid complications due to missing data points, only sites with a high data availability in wind speed, wind direction and surface pressure are chosen. This requirement limits the number of suitable sites compared with the error evaluation in Chap. 6. Hence, the investigation here will focus on four locations (see Fig. 5.1): Fehmarn and Hilkenbrook with 30-m wind data and Syke and Rapshagen with 10-m wind data. Fehmarn is an island in the Baltic Sea close to the German coast. The site that is investigated is near the island's shore line and, hence, exposed to rather inhomogeneous conditions. Northerly and north-westerly winds approach the site over the sea, while easterly to south-westerly winds arrive over land. In contrast to this, Hilkenbrook is in the north-western part of Germany about 70 km south of the coast of the North Sea. The terrain is rather homogenous in terms of the surface roughness and no orographic effects are to be expected. Syke is also located about 50 km east of Hilkenbrook in a slightly hilly terrain. This also holds for Rapshagen, which is situated in the east of Germany about 100 km south of the Baltic Sea. The results are presented in the following way. First the modes extracted from the meteorological variables with PCA are presented. Then cluster analysis is applied to the reduced data matrix, and the typical weather classes that are obtained by a straightforward approach are shown for complete as well as Ward's linkage and compared to large-scale weather maps. The average forecast error of the wind speed for each cluster is provided and discussed for the different linkage techniques.

9.3.1 Extraction of Climatological Modes

The PCA as described in Sect. 9.2.1 is carried out numerically using routines from the standard software MATLAB. In Fig. 9.1 the eigenvalues, λ_i, of the PCA for Fehmarn, Hilkenbrook and Rapshagen are shown. The eigenvalues are normalised with the total variance, i.e. the sum of all eigenvalues. It can be clearly seen in Fig. 9.1 that the eigenvalues decay rapidly with the first six eigenvalues already explaining about 90% of the variance.

The spectra are surprisingly similar though the stations are in regions with distinct weather regimes and different local conditions around the site. There are only minor deviations for the first and most important eigenvalues, i.e. the distribution of variance among the corresponding PC is rather consistent. The eigenvalues for Syke are not shown as they are as expected to be almost identical to those of the rather adjacent site Hilkenbrook.

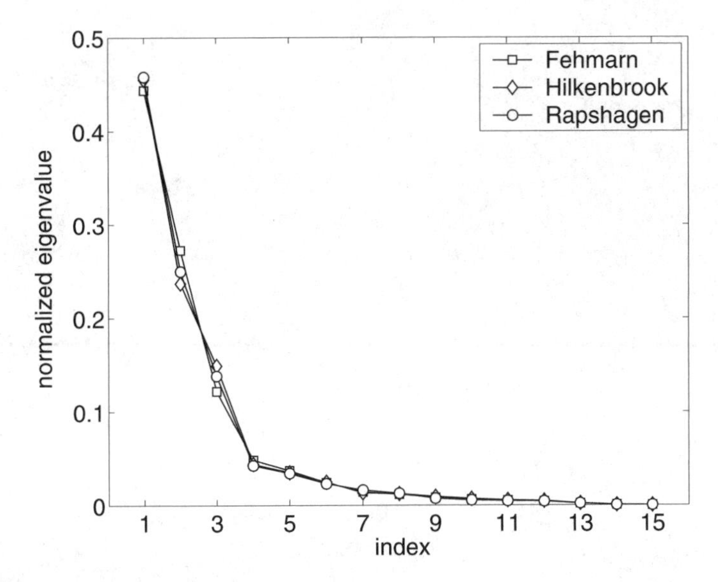

Fig. 9.1. PCA eigenvalue spectrum normalised with the sum of the eigenvalues, i.e. the total variance, for three sites (Fehmarn, Hilkenbrook and Rapshagen). Though the sites have different local wind characteristics, the spectra are similar

The first six principal components (PC) in the $\vec{u}$ space of Fehmarn and Hilkenbrook are shown in Fig. 9.2. The temporal evolution of $\vec{u}(t) = (u_t, v_t)$ at times $t = 0, 6, 12, 18$ and 24 h is indicated by connecting the end points (u_t, v_t) with lines. Note that the wind vector, i.e. the direction in which the wind blows, is shown, while the description of wind directions will refer to the direction from which the wind comes. Hence, e.g., southerly winds refer to wind vectors pointing to the north.

The corresponding PC of pmsl are illustrated in Fig. 9.3. Note that the PC do not necessarily correspond to actual meteorological conditions at the location on a specific day. They constitute the set of vectors whose linear combination can approximately reconstruct the current situation (cf. (9.4)).

Typically, the first two PC are rather stationary in that they show little diurnal variation in terms of wind direction and pressure. For all investigated sites the first PC describes moderate wind speeds from east or north-east, with slightly higher wind speeds at noon and virtually no change in wind direction over the day (see Figs. 9.2 and 9.3). In addition, the pmsl is at a constantly high level. PC 2 is also quite consistent for different locations. It refers to north-westerly wind directions with increasing wind speeds at midday. The corresponding pmsl is slightly rising at a high level. In terms of $\vec{u}$ these first two PC can roughly be associated with the two most frequent flow directions at the site.

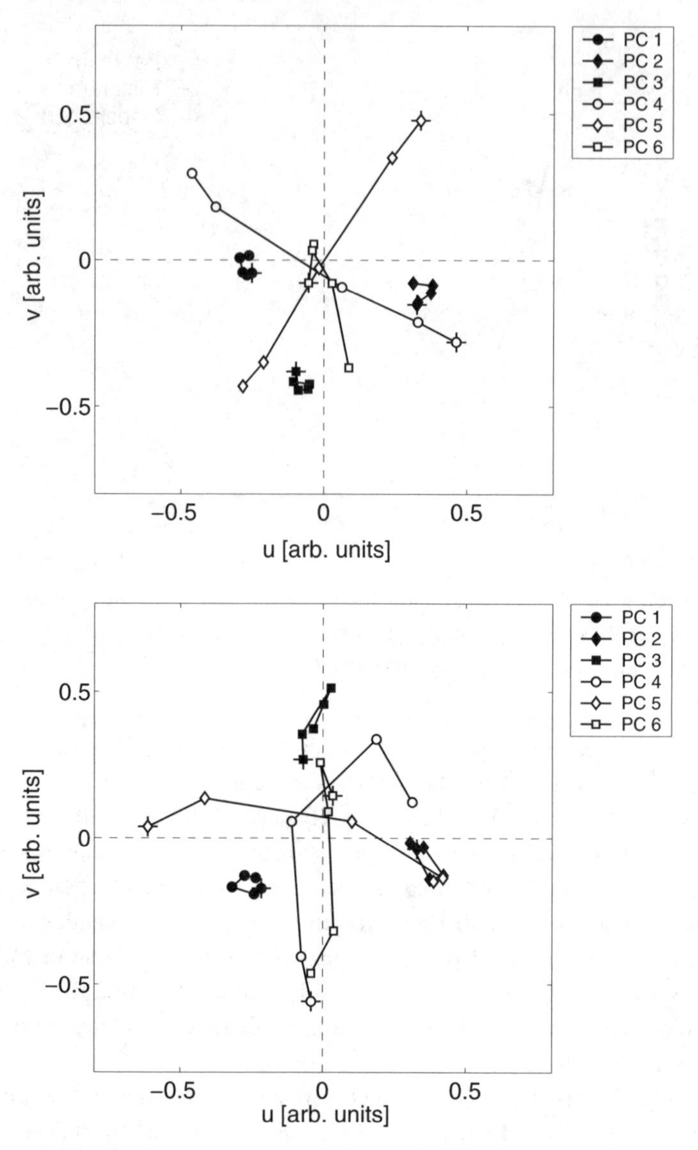

Fig. 9.2. First six principal components (PC) of $\vec{u}$ space for Fehmarn (*top*) and Hilkenbrook (*bottom*). The symbols denote the points (u_t, v_t) at times $t = 0, 6, 12, 18, 24$ h, where (u_0, v_0) ($t = 0$ h) is marked by "+". The first three PC are stationary, with slight wind speed variations over the day. The higher PC describe changes in the meteorological situation

The third PC is again similar for all sites although it occurs with different signs. As the sign of the PC is arbitrary, those PC which only differ in their sign are regarded as identical climatological modes. In half of the cases PC 3 refers to wind from the north, with pmsl increasing from low to medium level, while for the other half it

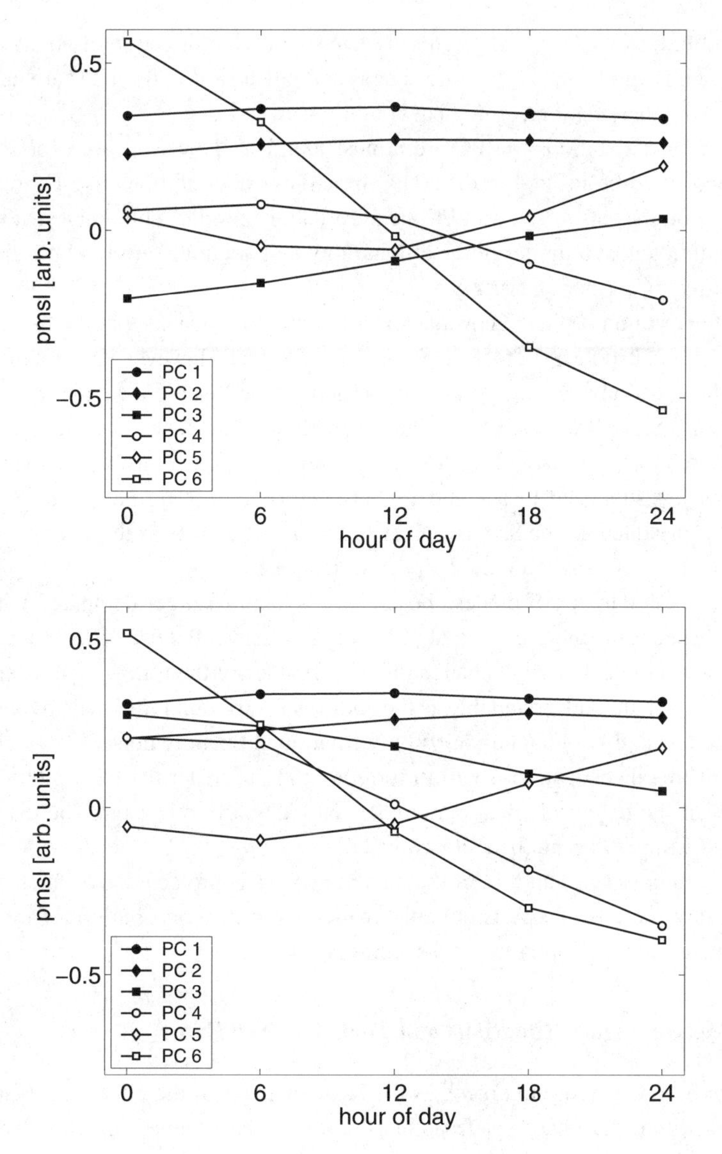

Fig. 9.3. First six PC of pmsl space for Fehmarn (*top*) and Hilkenbrook (*bottom*). The first two PC refer to rather constant high pressure, while the higher PC describe changes in pmsl which can be rather dramatic, e.g. PC 6

describes wind from the south in connection with pmsl falling from high to medium level.

Higher principal components are much more dynamic for all sites than the first three PC. As can be seen in Figs. 9.2 and 9.3 the diurnal variations in all variables are profound. PC 4–6 typically refer to changes in wind direction of about 180° up to 300° with rapidly changing wind speeds and pmsl. Hence, these PC are needed to account for changing meteorological conditions over 1 day.

For adjacent stations the PC are almost identical. However, even for sites that are supposed to be located in different climatological conditions like Fehmarn and Hilkenbrook the sets of relevant PC are surprisingly similar. This suggests that PCA extracts fundamental modes of the climatology that are quite universal for the investigated area of northern Germany.

The inspection of the eigenvalue spectra in Fig. 9.1 reveals that the first four PC on average contribute most to the meteorological signal. However, the structure of the PC as discussed above showed that, in particular, the PC 4–6 describe the dynamic changes of the weather condition occurring within 24 h. Hence, these higher modes are expected to represent more extreme meteorological situations which do not occur often but are supposed to account for larger forecast errors than more stationary modes. Consequently, the first six PC are chosen, i.e. $N = 6$, as the orthogonal basis $\{\vec{q}_i\}$, $i = 1, \ldots, 6$, spanning the reduced state space.

Hence, the number of modes chosen here is slightly larger compared with what the eigenspectrum suggests. Yarnal [124] points out that the use of too many eigenmodes does not necessarily enhance the performance of the following cluster analysis, which is quite understandable as the additional variation introduced by including more modes might not contain useful information but mainly noise. However, in the course of this investigation it will turn out that in particular the PC which describe changes in the meteorological variables over 1 day are important. The use of less than six PC has so far not been investigated.

According to (9.2) and (9.3), the data matrix M is projected onto the new basis and the day-score matrix X is obtained that contains the contributions of the six PC to the meteorological situation of a specific day.

9.3.2 Meteorological Situations and Their Forecast Error

The 365 by 6 day-score matrix X is the basic element of the cluster analysis as its rows constitute the phase space points that are to be grouped together. The three linkage techniques average, complete and Ward's linkage have been used in this investigation to define clusters. In contrast to the linkage provided by Shagedanova et al. [103] and Kalkstein et al. [51], average linkage as defined in (9.7) failed in producing meaningful synoptic classes as it sorted two thirds of the days into one common group and divided the remaining days into a number of small groups. Complete linkage (9.6) and Ward's linkage (9.8) produce clusters that can be associated with typical weather situations, and, hence, the classification schemes based on these linkage types appear to be reasonable. In the following the results of complete and Ward's linkage are presented for the two sites Fehmarn and Hilkenbrook.

Complete Linkage

The first step of the cluster analysis is to successively join the objects in the phase space starting from 365 separate days up to one single cluster and record the distances, according to (9.6), of the two clusters that have been formed in each step. The larger the distance, the greater the difference between the clusters in terms of the chosen linkage method. Hence, the clustering process should be interrupted if the distances between the clusters become too large, which is often indicated by a "jump" in the series of distances.

Figures 9.4 and 9.5 show the distances of the final 15 clustering steps using complete linkage at the two sites. In both cases there are several jumps in the distances that suggest the unification of two clusters that are quite distinct. For Fehmarn, two crucial points can be identified at the transition from seven to six and from five to four clusters. At Hilkenbrook, detectable steps in the cluster distances occur at the transition from eight to seven and also five to four clusters.

In this investigation it turns out that the classification scheme is quite robust in the sense that useful classification results can be obtained for different cluster numbers. The behaviour of the cluster distances suggests a range of possible cluster numbers that should be considered for further investigations rather than providing just one optimal number. Which number of clusters is selected depends on the detailed purpose

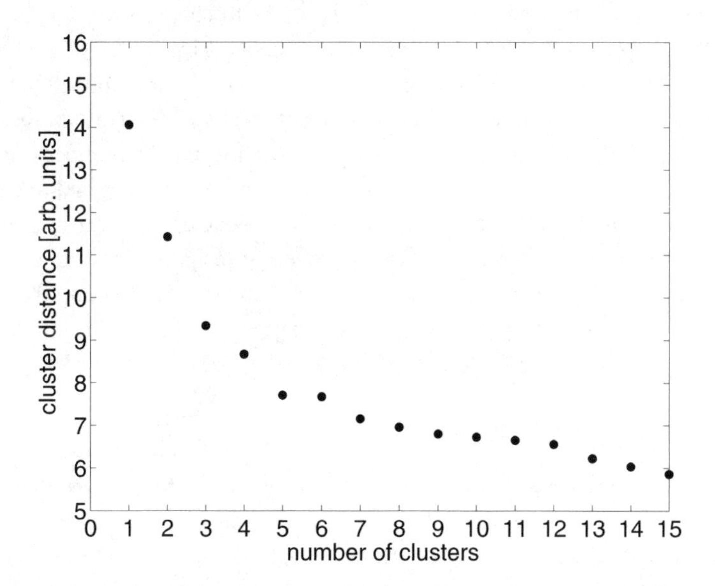

Fig. 9.4. Distance between clusters versus number of clusters using complete linkage at Fehmarn. As the clusters are successively joined, "jumps" in the distances start to occur at the transitions from seven to six and five to four clusters. This defines the range of cluster numbers considered for further analysis

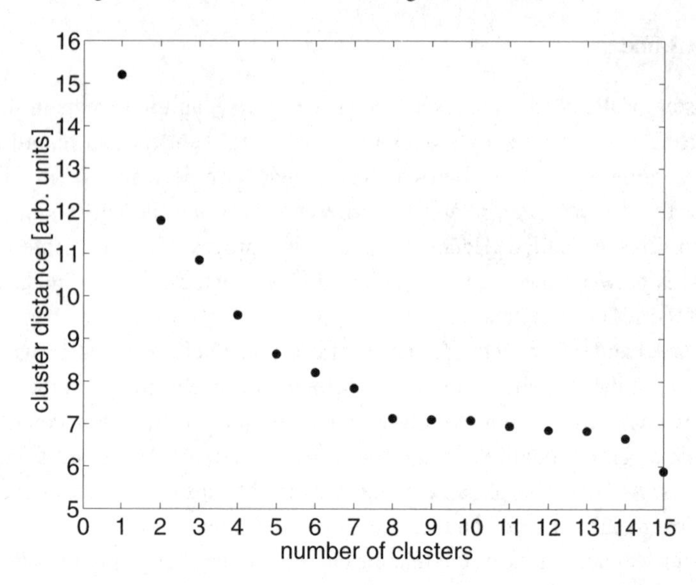

Fig. 9.5. Distance between clusters versus number of clusters using complete linkage at Hilkenbrook. Note the discontinuities at the transitions from eight to seven and five to four

of the investigation. In order to interpret clusters in terms of weather classes and to compare clusters found at different sites, it is useful to choose six to eight clusters because the synoptic situations are more homogeneous. However, using more clusters means that a smaller proportion of the 365 days per year are grouped together in a common cluster. This can lead to less significant results with regard to the average forecast error. As a consequence, it seems advisable to vary the number of clusters in the range indicated by the discontinuities in the cluster distances and to carefully consider the results in terms of the corresponding weather class on the one hand and the forecast error on the other. As the clustering technique has the convenient property that only two clusters are joined at each step without affecting the remaining groups, it is easy two follow what actually happens if the number of clusters is varied.

Complete Linkage at Fehmarn

Following this approach six clusters are used for Fehmarn to see how they can be connected to the overall weather situation. After the clusters are defined the corresponding meteorological situation is considered in terms of the mean values of "real" $\vec{u}$ and pmsl for each cluster. Fig. 9.6 shows the mean values of the meteorological variables of Fehmarn's six clusters. Note that $\vec{u}$ denotes the wind vector and points in the direction in which the wind blows, whereas wind directions, as usual in meteorology, refer to the direction from which the wind blows. In terms of pmsl (Fig. 9.6

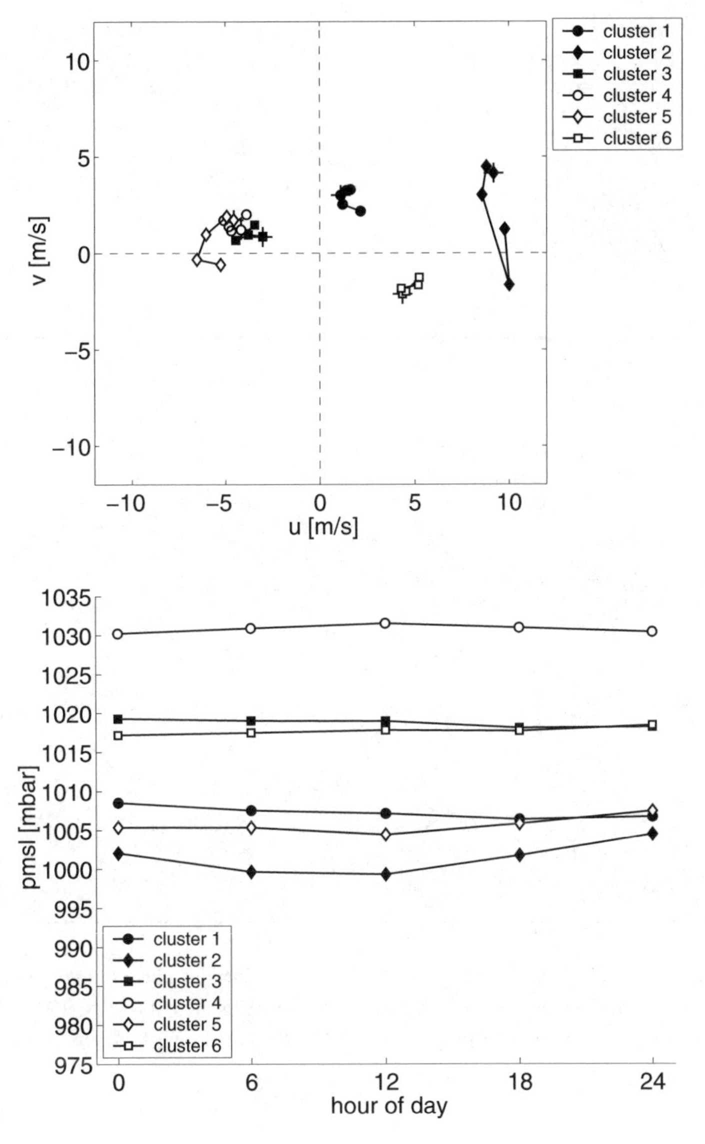

Fig. 9.6. Means of $\vec{u}$ (*top*) and pmsl (*bottom*) for different forecast horizons for six clusters at Fehmarn constructed with complete linkage. For $\vec{u}$ the symbols denote the points (u_t, v_t) at times $t = 0, 6, 12, 18, 24$ h, where (u_0, v_0) $(t = 0$ h) is marked by "+". The corresponding weather situations are shown in Figs. 9.7, 9.8 and 9.9

(bottom)) the clusters show three different regimes: low pressure (clusters 1, 2 and 5), moderately high pmsl (clusters 3 and 6) and very high pmsl (cluster 4).

The low-pressure situations differ considerably with respect to wind speed and direction (Fig. 9.6 (top)). Typically, cluster 1 has moderate winds from south-west

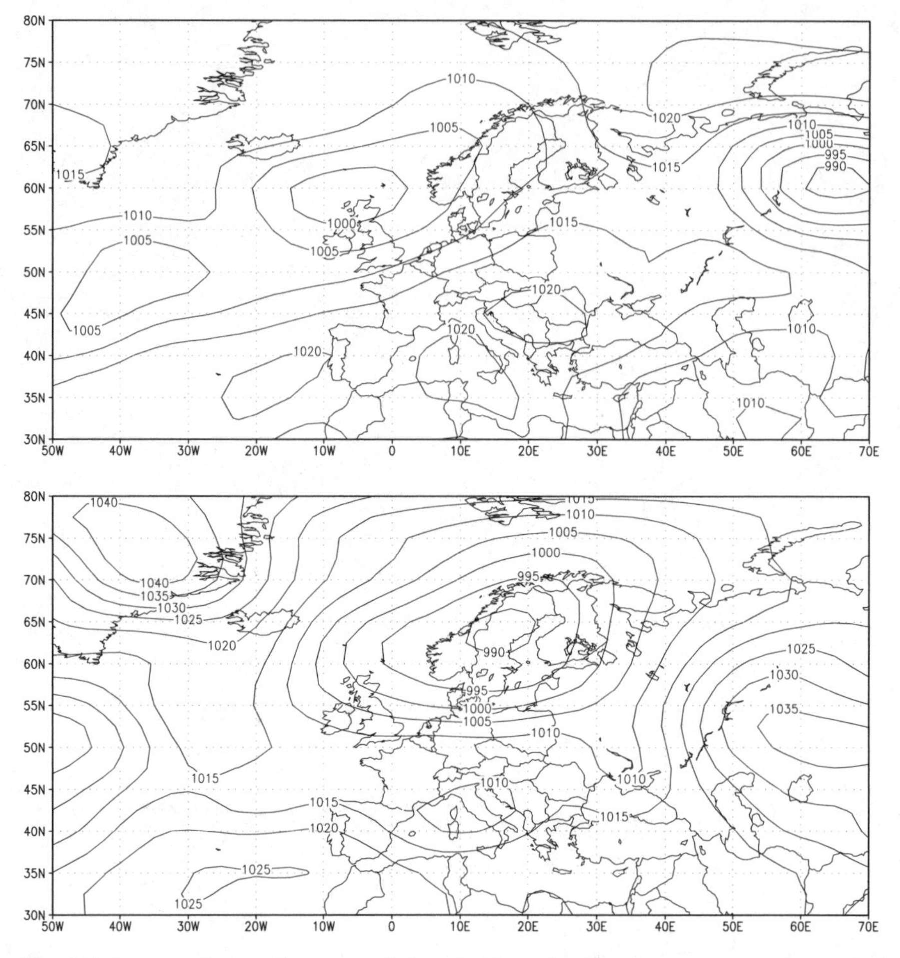

Fig. 9.7. Large-scale weather maps of days inside a specific cluster representing typical weather classes derived by complete linkage at Fehmarn located at 54° N, 12° E. *Top*: Low pressure approaching from north-west (cluster 1). *Bottom*: Strong low passes north of Germany (cluster 2)

and slightly decreasing pressure caused by a low-pressure system approaching from the west or north-west together with a high south of the site. The large-scale weather map of a typical day of cluster 1 is shown in Fig. 9.7 (top). Cluster 2 refers to a strong low that passes north of the site, as illustrated in Fig. 9.7 (bottom). The pmsl drops on average to a low level around 1000 mbar and recovers again at the end of the day. The wind speeds are very high with wind direction changing from south-west to west, north-west and then back to south-west. Cluster 5 is related to a situation where the low passes west of the site, e.g. over Britain and France, with high pressure gradients

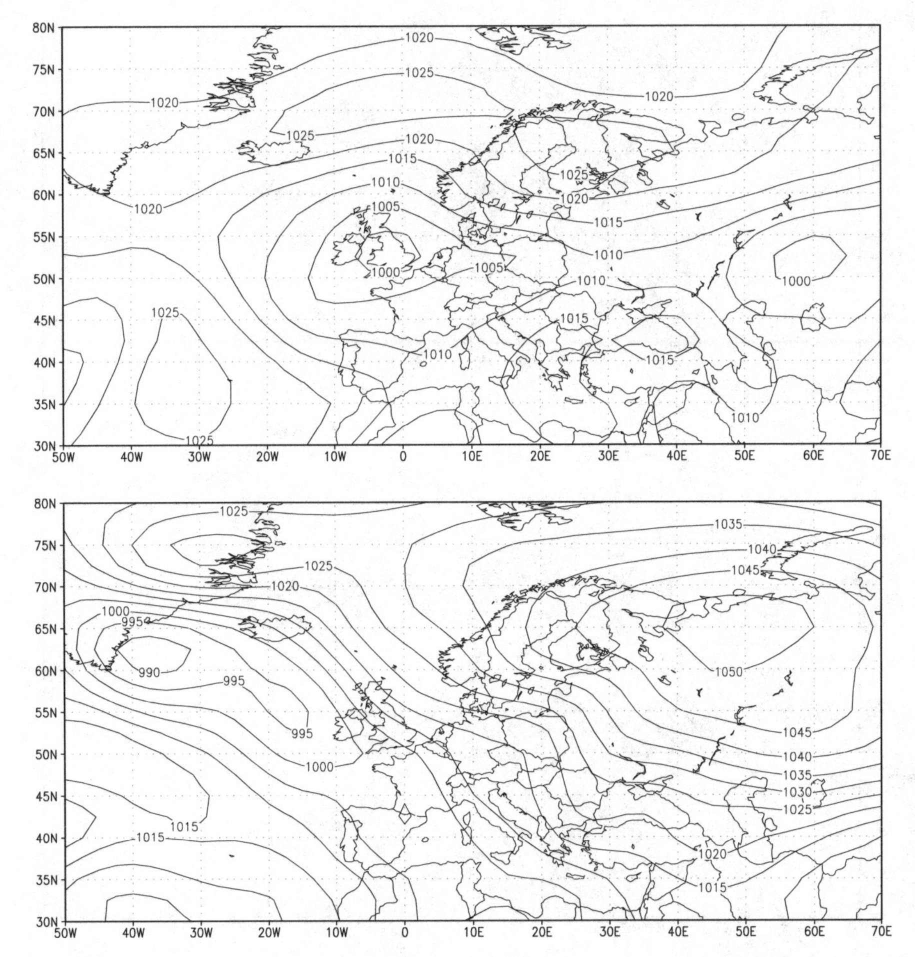

Fig. 9.8. Large-scale weather maps of days inside a specific cluster representing weather classes different from those in Fig 9.7 derived by complete linkage at Fehmarn located at 54° N, 12° E. *Top*: Low passing west of Fehmarn over Britain and France (cluster 5). *Bottom*: Stationary high-pressure over western Russia (cluster 3)

over the Baltic Sea (Fig. 9.8 (top)). Wind speeds are quite considerable turning from south-east to north-east.

Cluster 3 is related to moderate wind speeds from easterly directions. This typically occurs if a stable high-pressure area persists over western Russia and a rather strong low is located west of the site, leading to considerable pressure gradients over Fehmarn (Fig. 9.8 (bottom)).

Days in cluster 6 are characterised by almost the same mean pressure as in cluster 3 but with north-westerly wind directions. In this case Central Europe is influenced by a high or ridge located west or north-west of the site (Fig. 9.9 (top)).

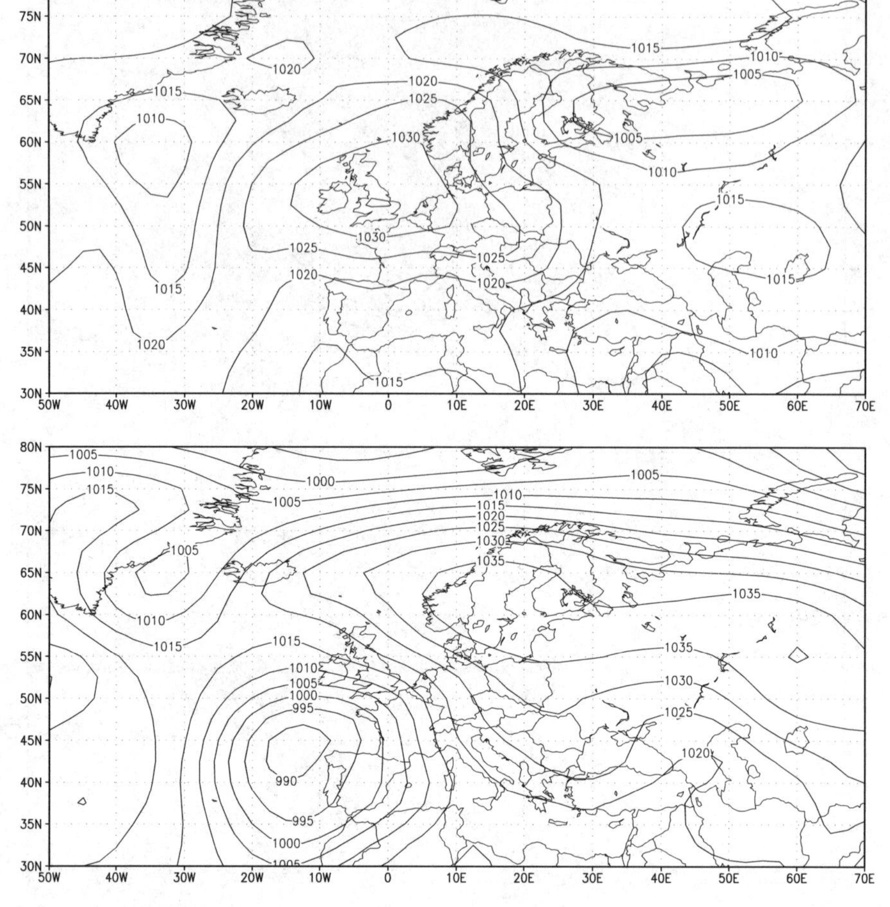

Fig. 9.9. Large-scale weather maps of days inside a specific cluster representing weather classes different from those in Fig. 9.7 derived by complete linkage at Fehmarn located at 54° N, 12° E. *Top*: High-pressure area north-west of Germany, with low over western Russia (cluster 6). *Bottom*: Strong high-pressure system over Scandinavia, with low influencing west of Europe (cluster 4)

Cluster 4 summarises mainly days from the winter period with very high pmsl around 1030 mbar and easterly winds. The corresponding weather situation is typically associated with high-pressure over Scandinavia extending over Central Europe as shown in Fig. 9.9 (bottom).

Comparing the large-scale weather maps of days within one cluster shows that the overall weather situation for most of the days is rather coherent. Hence, the classification scheme that is based on local data at the site can be related to the overall weather situation. This is not totally surprising as the local meteorological situation described by wind speed, direction and pressure is caused by the atmospheric circulation on a larger scale. Of course, for some days the weather situations are not very precisely described by the mean values of the cluster. If more classes are used, clusters typically split into new clusters which are in itself more homogeneous than before. But with regard to the investigation in this work a small number of clusters is desired and sufficient even though the division might be too rough for other purposes. On the whole, the synoptic classification using six clusters with complete linkage for Fehmarn appears to be reasonable.

Complete Linkage at Hilkenbrook

Hilkenbrook is about 250 km west of Fehmarn and, therefore, expected to be exposed to different wind conditions, in particular, less influenced by continental high-pressure systems over Scandinavia and Russia but more affected by lows approaching from the Atlantic Ocean. However, there should be a number of weather situations which are comparable to Fehmarn as the size of the overall circulation patters is much larger than the distance between the sites. This is confirmed by considering the means of the meteorological variables in the seven clusters created by complete linkage for Hilkenbrook in Fig. 9.10. Again the clusters are roughly ordered according to low pressure (clusters 5 and 6), moderate pmsl (clusters 3 and 4), and high pmsl (clusters 1, 2 and 7).

The two low-pressure classes both refer to rather dynamic situations, with pmsl varying over the day (Fig. 9.10). Cluster 5 has high wind speeds from the south, with pmsl dropping from about 1005 mbar to 995 mbar due to a low-pressure system approaching from the north or north-west. Cluster 6 is related to a strong low passing north of the site. The wind blows rather strongly and turns from south-west to west and back to south-west indicating the passage of a front.

Moderate pmsl of around 1010 mbar occurs at Hilkenbrook mainly in two different weather regimes. Cluster 3 shows rather high wind speeds from the east and slightly increasing pressure typically related to a high-pressure system east or north-east of the site, e.g. over western Russia, and low-pressure in the west or south-west. Cluster 4 contains quite a number of days which have pmsl of around 1010 mbar and very low wind speeds. The weather class is not as clear as in all the other clusters but mainly characterised by the fact that the centres of the pressure areas are far away, with small pressure gradients at the location of the site.

The three high-pressure clusters at Hilkenbrook show quite pronounced differences concerning their mean wind directions. Cluster 1 has on average the highest pmsl of about 1023 mbar. Note that this is more than 5 mbar lower compared with the highest pmsl at Fehmarn. The wind directions are from the east with moderate speeds. The overall weather situation is typically dominated by a stationary high-pressure system over Scandinavia and, hence, north-east of the site. In contrast to

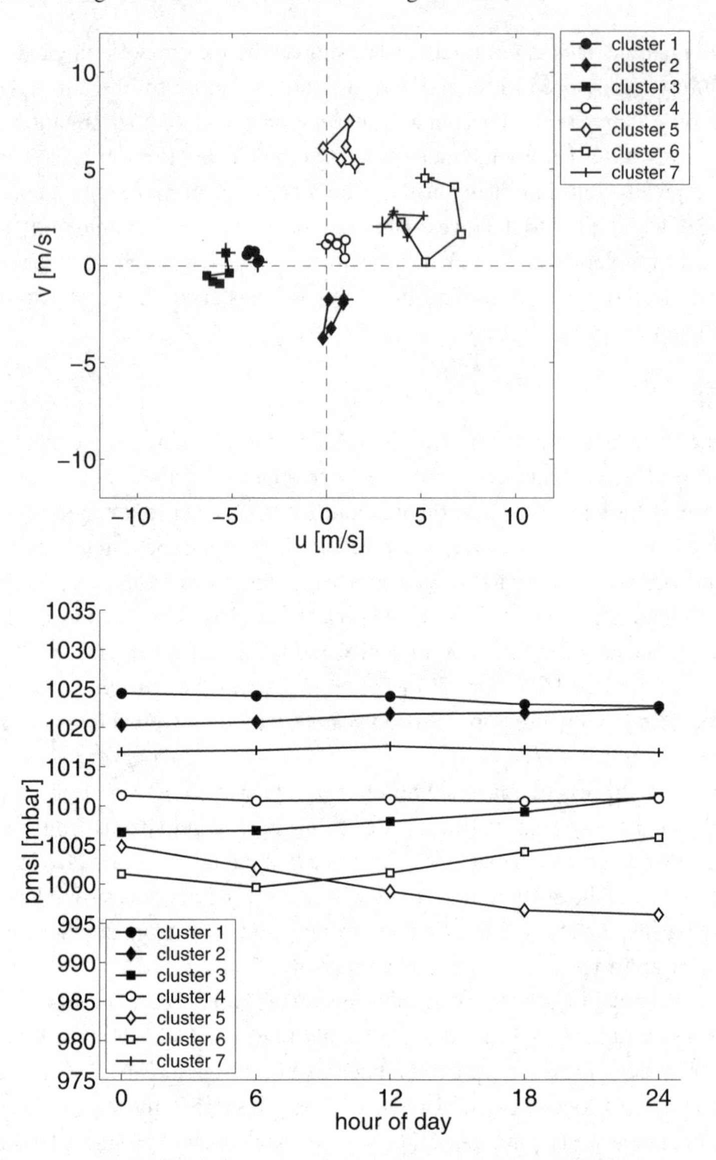

Fig. 9.10. Means of $\vec{u}$ (*top*) and pmsl (*bottom*) at different prediction times for seven clusters found for Hilkenbrook with complete linkage. For $\vec{u}$ the symbols denote the points (u_t, v_t) at times $t = 0, 6, 12, 18, 24$ h, where (u_0, v_0) $(t = 0$ h$)$ is marked by "+"

this, cluster 2 refers to a weather situation where the high-pressure area is located in the west of the site, leading to northerly wind directions. Finally, cluster 7 is related to south-westerly winds caused by high-pressure south-west or south of the site and at the same time low-pressure in the north.

Hence, Hilkenbrook also, for the classification of meteorological situations based on complete linkage appears to be useful.

The typical weather classes found for the two sites seem to be rather similar in terms of the general description of the meteorological situations. In order to further test the consistency of the classification scheme the days that simultaneously appear in clusters from Fehmarn and Hilkenbrook are counted. The result is shown in Table 9.1, where the rows refer to Fehmarn's clusters and the columns to those from Hilkenbrook.

Table 9.1. Comparison of equal days in clusters for Fehmarn and Hilkenbrook

	Hilkenbrook							
	Cluster 1	Cluster 2	Cluster 3	Cluster 4	Cluster 5	Cluster 6	Cluster 7	$\sum$
Fehmarn								
Cluster 1	0	0	0	29	12	6	3	50
Cluster 2	0	0	0	0	0	15	2	17
Cluster 3	30	16	15	51	2	0	7	121
Cluster 4	35	0	0	0	0	0	0	35
Cluster 5	0	0	3	16	1	1	0	21
Cluster 6	4	43	0	38	0	0	36	121
$\sum$	69	59	18	134	15	22	48	365

Some clusters share a rather profound number of days. For example, cluster 2 of Fehmarn is, except for 2 days, completely contained in Hilkenbrook's cluster 6. Hence, on these 15 days the classification at both sites records a weather situation dominated by a strong low-pressure system that is centred north of Germany and moves quickly to the north-east. However, the exact path and the advection time of the low decides whether both sites are affected simultaneously or only one of them. If the low takes a more southern route, only Hilkenbrook classifies this situation as cluster 6 while Fehmarn has a smaller pressure drop and lower wind speeds as it is located further in the east and, hence, might not be affected by the fronts. Fehmarn labels this situation as cluster 1.

Another example is cluster 4 of Fehmarn, which is totally absorbed by cluster 1 of Hilkenbrook. These 35 days are typically related to a high-pressure system over Scandinavia or western Russia, with extremely high pmsl over north-eastern Germany. However, Hilkenbrook's cluster 1 contains more days than that which overlaps with cluster 3 of Fehmarn. At these days the high-pressure system is located further in the east, leading to similar wind directions from the east and pmsl of around 1020 mbar at both sites.

On the other hand, certain weather situations lead to different classification results at both stations, e.g. cluster 3 of Fehmarn is distributed over six different clusters of Hilkenbrook. This occurs if the two sites are influenced by distinct weather regimes. In this case Fehmarn is dominated by a rather high-pressure located over western Russia, while at the same time Hilkenbrook can experience a variety of high- or lower pressure situations.

The result of this comparison is that the consistency between the classifications at the two sites seems to be rather high. For the majority of cases, common days in clusters can be plausibly explained by the overall weather situation that either shows that the two sites are affected in the same way or why they simultaneously record different local conditions.

Relation to Forecast Error Using Complete Linkage

For each of the clusters the mean forecast error is determined by averaging the rmse values calculated for each day with (9.9). As described in Chap. 5 the wind speed prediction at 10 m is provided by the "Deutschlandmodell" of the German Weather Service, while the corresponding wind speed measurements are from the WMEP programme. Thus, the errors are determined based on the same data that have been used in Chap. 6 to evaluate the overall performance of the wind speed forecast.

The means of the daily rmse values per cluster normalised to the annual average of the wind speed are shown in Fig. 9.11 for Fehmarn and Fig. 9.12 for Hilkenbrook found with complete linkage. In both cases there are considerable differences between the forecast errors of the different clusters. The error bars illustrate the 95% confidence intervals of the mean values suggesting that the clusters with minimum and maximum rmse are indeed significantly separated (detailed results of the statistical tests on this matter can be found in Appendix B).

In particular, Fehmarn's clusters 1, 2 and 5 are related to large relative rmse values of around 0.38 corresponding to about 2.1 m/s absolute rmse. As described before (Fig. 9.6), these three clusters are related to different low-pressure situations. Cluster 2 corresponds to situations where a strong low passes north of the site and has on average the highest forecast error. In contrast to this, cluster 4 representing the weather type with the largest pressure has the smallest rmse of about 0.23 relative and 1.3 m/s absolute. A statistical F test (with confidence level 0.05) confirms that this is significantly lower compared with the above-mentioned classes 1, 2 and 5 (see Table B.2). The ratio between the largest and the smallest rmse is 1.7, which is very profound.

At Hilkenbrook cluster 6 has the highest rmse with 1.8 m/s absolute and 0.51 relative. The corresponding weather situation is related to a low-pressure system passing north of the site. The cluster with the smallest average forecast error is cluster 1 related to a rather stationary high-pressure situation where the rmse is 1.0 m/s

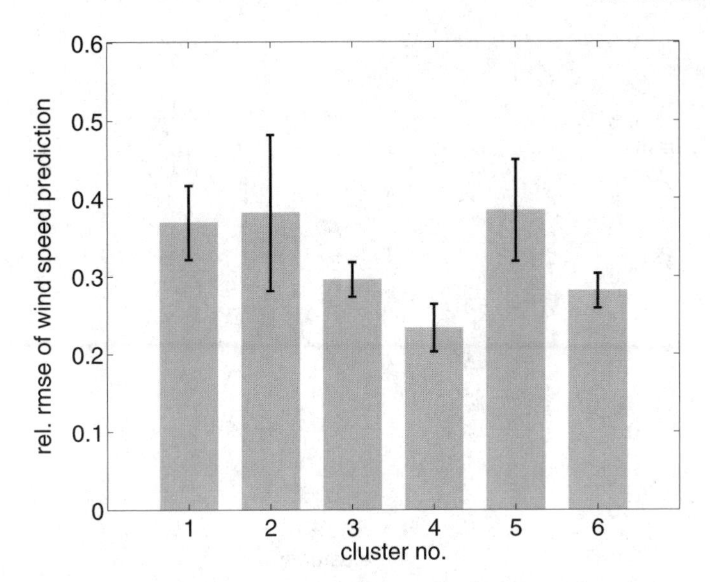

Fig. 9.11. Means of daily rmse of wind speed prediction normalised with annual mean of wind speed for the six clusters found with complete linkage at the site Fehmarn. The error bars illustrate the 95% confidence intervals of the mean. Clusters 1, 2 and 5 show the largest error. In both cases the corresponding weather situation is dominated by a low-pressure system. All three clusters have significantly larger forecast errors than cluster 4 which typically refers to a stationary high-pressure situation and has a significantly lower rmse compared with clusters 1, 2 and 5 (see Table B.2)

absolute and 0.29 relative to the mean wind speed. The ratio between the maximum and minimum rmse is 1.8. It is also interesting to note that the forecast error in cluster 4 is on a medium level of 0.39 relative rmse (1.4 m/s), which is roughly halfway between the smallest and largest error values and significantly different from both of them. The days in this cluster are related to medium pmsl with small pressure gradients and, hence, low wind speeds.

Again the average forecast errors found for the typical weather situations at the two sites show a good consistency. For Fehmarn as well as for Hilkenbrook, large rmse values occur for the clusters that are related to a rather fast-moving low-pressure system that passes north of the site (cluster 2 for Fehmarn, 6 for Hilkenbrook). The local meteorological conditions are characterised by pmsl on a relatively low level decreasing further around midday but recovering at the end of the day. Wind speeds are fairly high with wind directions turning from south-west to west and back to south-west, indicating the passage of a front. Depending on the path of the low, this situation simultaneously occurs at both sites or only at Hilkenbrook (see Table 9.1).

In addition, for Fehmarn the passage of low-pressure areas in the west or south-west of the site with strong easterly winds (cluster 5) is also related to large forecast

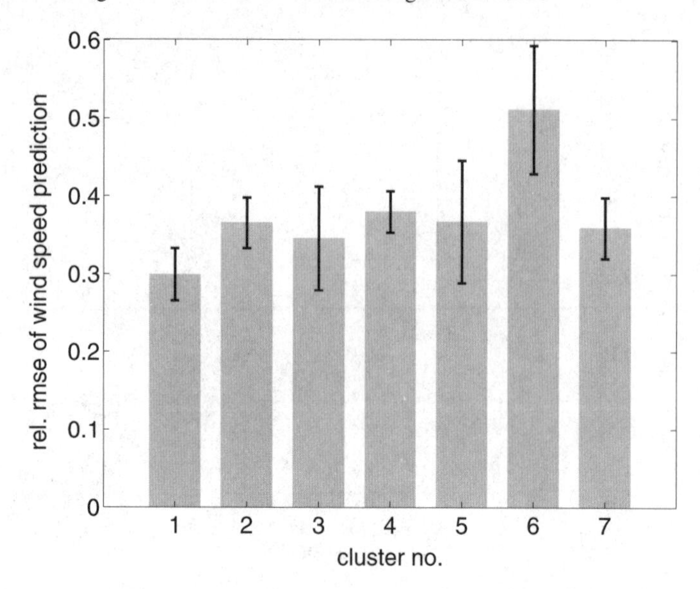

Fig. 9.12. Means of daily rmse of wind speed prediction normalised with annual mean of wind speed for the seven clusters found with complete linkage at Hilkenbrook. Cluster 6 has the maximum forecast error, which is significantly different from clusters 1, 2, 4 and 7. Cluster 6 is related to a low-pressure system passing north of the site. Cluster 1 with the smallest rmse corresponds to a situation where a stationary high pressure dominates the local weather conditions. Its error is significantly lower compared with clusters 4 and 6 (see Table B.4)

errors. At Hilkenbrook these days are recorded in cluster 4, which additionally contains more situations with smaller wind speeds on average and slightly higher pmsl compared with Fehmarn. The corresponding forecast error is the second largest at Hilkenbrook but significantly lower than the largest one. Thus, at Fehmarn there are higher wind speeds caused by larger pressure gradients, while at the same time at Hilkenbrook more moderate conditions prevail.

In contrast to this, both sites have the smallest forecast error in situations where a high-pressure area lies rather stationary over Scandinavia or the Baltic Sea, with relatively high wind speeds from the east. For Fehmarn the mean rmse in this case is 17% to 39% smaller than that of the other weather classes. For Hilkenbrook the minimum rmse is 13% to 42% smaller than in the other clusters.

Hence, using complete linkage leads to a reasonable synoptic classification of the meteorological data (30-m wind data and pmsl of nearby synop station) at these two sites in the sense that the clusters can be associated with typical large-scale weather classes. The number of classes found by considering the interval of clustering steps given by "jumps" in the distances between the clusters as criterion to stop the clustering process already provides a good choice. Moreover, the mean forecast errors related to the clusters show significant differences where the maximum and minimum rmse are, as expected, related to dynamic low-pressure and stationary high-pressure, respectively.

Ward's Linkage

For most of the investigated sites Ward's linkage produced clusters which are in principle comparable to those found by complete linkage. But the alterations between the groups of days generated by the two methods can make important differences in terms of the grouping of days and, thus, the related forecast errors. This is now illustrated for Fehmarn, where typical differences between the methods occur.

As Fig. 9.13 shows, the distances between the clusters at the final 15 steps of the clustering procedure behave rather similar to complete linkage. There are discontinuities at the transitions from eight to seven and five to four clusters. Again, choosing a number of clusters from this interval leads to a classification that allows to relate the clusters to large-scale weather situations and to find significant differences between the average forecast errors per cluster.

The results of the classification based on six clusters with Ward's linkage for Fehmarn are shown in Fig. 9.14. With regard to these mean values of $\vec{u}$ and pmsl, the classes are generally comparable to those found by complete linkage in Fig. 9.6. But it is obvious by comparing the means of the pressure that there are differences in par-

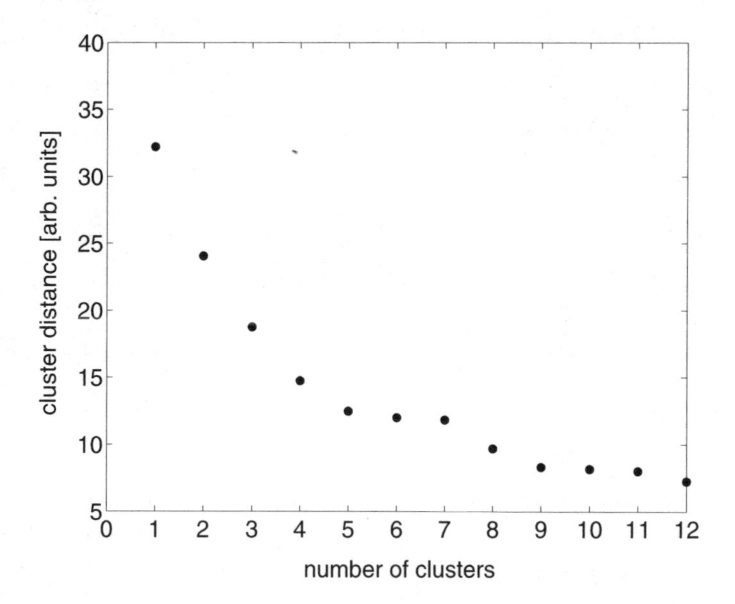

Fig. 9.13. Distance between clusters versus number of clusters using Ward's linkage at Fehmarn. "Jumps" in the distances occur at the transitions from eight to seven and five to four clusters. This is similar to complete linkage (Fig. 9.4)

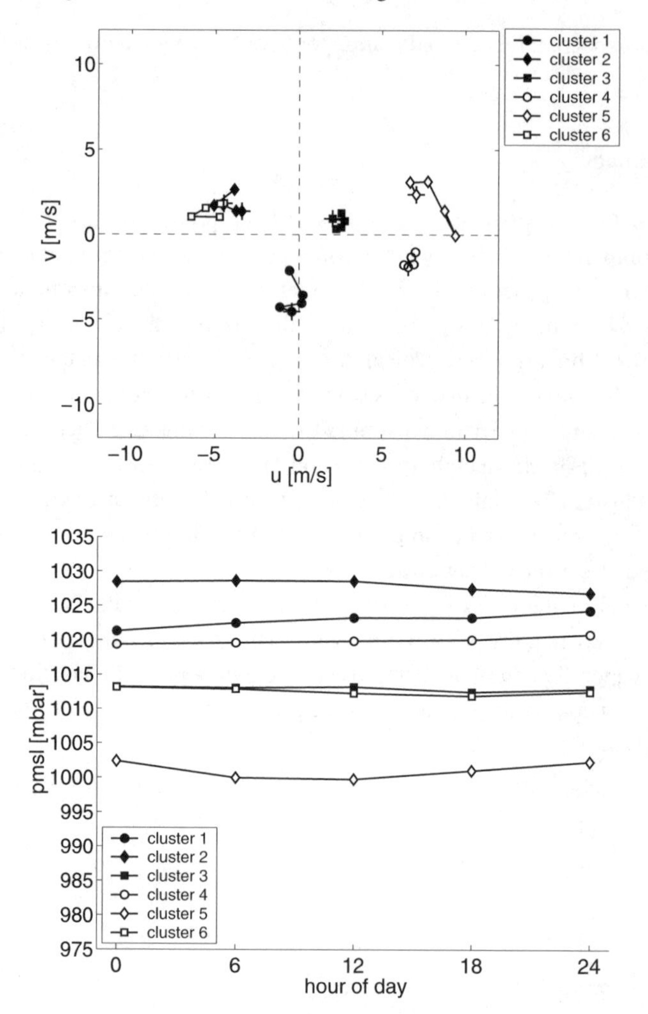

Fig. 9.14. Means of $\vec{u}$ (*top*) and pmsl (*bottom*) at different forecast horizons for six clusters found by Ward's linkage at Fehmarn. For $\vec{u}$ the symbols denote the points (u_t, v_t) at times t = 0, 6, 12, 18, 24 h, where (u_0, v_0) (t = 0 h) is marked by "+". Some of the clusters are comparable with those provided by complete linkage (Fig. 9.6), in particular cluster 5 with cluster 2 by complete linkage

ticular for the low to medium range, with pmsl between 1005 mbar and 1015 mbar. Moreover, in terms of the mean wind vectors, Ward's cluster 1 appears to constitute a new class which does not appear under complete linkage.

A more detailed comparison of the overlaps between the clusters reveals that some days are grouped rather differently by the two methods. Table 9.2 shows that the only two clusters that are nearly identical are number 2 of complete and 5 of

Table 9.2. Comparison of equal days in clusters using complete and Ward's linkage for Fehmarn[a]

	Ward						
	Cluster 1	Cluster 2	Cluster 3	Cluster 4	Cluster 5	Cluster 6	$\sum$
Complete							
Cluster 1	0	0	30	0	4	16	50
Cluster 2	0	0	1	0	16	0	17
Cluster 3	10	24	32	0	0	55	121
Cluster 4	0	33	0	2	0	0	35
Cluster 5	0	0	0	0	1	20	21
Cluster 6	22	0	41	55	3	0	121
$\sum$	32	57	104	57	24	91	365

[a]Though the means of the clusters are rather similar to complete linkage, there are differences in the actual grouping

Ward's, which share 16 common days related to the well-known weather situation of the passing low in the north. Hence, these seldom but rather extreme days are consistently classified by both methods.

Another interesting weather class is the high over Scandinavia or the Baltic Sea detected by complete linkage as cluster 4. Under Ward's linkage these 33 days with very high-pressure are almost totally contained in cluster 2. But this cluster contains additional days from situations with slightly lower pmsl and lower wind speeds. Hence, Ward's cluster 2 joins days into one class that are separately classified under complete linkage, i.e. clusters 3 (pmsl of around 1020 mbar) and 4 (very high pmsl $\approx$ 1030 mbar). This tendency of Ward's linkage to group extreme together with less extreme situations though they should remain distinct has already been described in earlier investigations, e.g. by Kalkstein et al. [51]. Another typical feature of Ward's method is to create rather equally sized groups, which can also be observed in Table 9.2. The distribution of days on the clusters is more balanced compared with complete linkage.

Relation to Forecast Error Using Ward's Linkage

The mean daily forecast error per cluster for Ward's linkage is shown in Fig. 9.15. Compared with the rmse of the complete linkage clusters in Fig. 9.11 the differences between the classes are less pronounced. The statistical F test reveals that cluster 2 with the lowest rmse is significantly different from clusters 3 and 6 with the highest rmse (cf. Table B.3). But in contrast to this, complete linkage results in a higher number of significantly different clusters; in particular the cluster with the smallest prediction error has a significantly lower rmse than three of the other clusters. Thus,

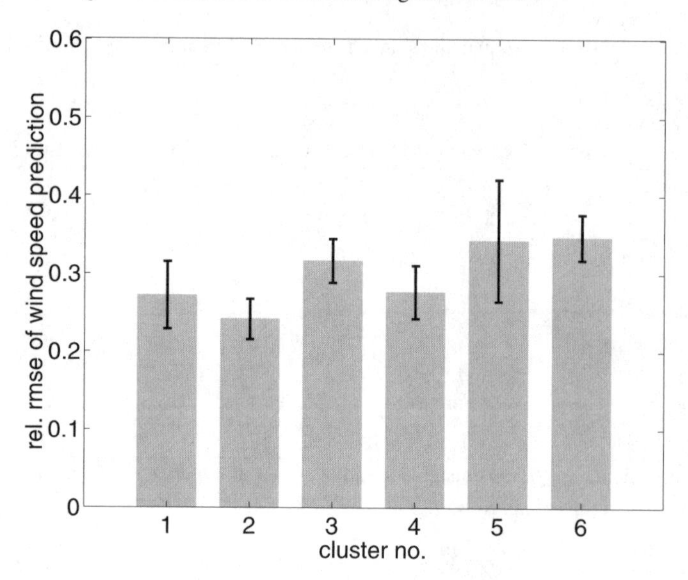

Fig. 9.15. Means of daily rmse for the six clusters found with Ward's linkage at Fehmarn. The differences between the forecast errors are not as pronounced as for complete linkage (Fig. 9.11 and Table B.3).

at this site Ward's clusters have rather equalised error levels with smaller differences between the clusters. This is considered as a disadvantage compared with complete linkage as the classification scheme that provides a better distinction between the forecast errors is more useful—of course, under the condition that it produces rather homogeneous clusters which can be associated with certain weather types.

For the sites in this investigation the clusters constructed by Ward's linkage provided a classification scheme that appears to correspond to typical weather situations. But compared with complete linkage the grouping of days can be different, with a slight trend to join distinct meteorological situations into common classes and to form equally sized groups. However, in contrast to the findings by Kalkstein et al. [51], Ward's linkage does not appear "to blur distinctions between types". In terms of the related prediction error, complete linkage performs better than Ward's linkage at Fehmarn and Hilkenbrook as it provides sharper distinctions between the clusters.

Results for Other Sites

The site Syke is about 50 km west of Hilkenbrook still in the north-western part of Germany. The results based on complete linkage using of measurements of wind data at 10-m height instead of 30 m as before are rather similar to those for Hilkenbrook. Here the passage of the low centred in the north is also related to the maximum forecast error. The minimum rmse also occurs for a high-pressure situation, but in contrast to the two sites, the high is west of the site. The Scandinavian high that leads to the minimum forecast errors for the sites still has a relatively small rmse. At Syke the performance of Ward's technique is comparable to the complete method. The behaviour of the forecast error for the different weather classes, particularly, the systematically larger rmse for dynamic low-pressure situations, is also confirmed under Ward's linkage.

At Rapshagen complete linkage (also 10-m wind data) provides clusters whose interpretation in terms of large-scale weather classes is not as clear as for the other sites. At this site Ward's linkage seems to be superior and leads to a classification that is easier to relate to weather patterns. The forecast errors per cluster again show that the well-known low-pressure passage has a rather high rmse, which certain high-pressure conditions also have. However, this site should be treated with care as local effects seem to strongly influence the flow conditions locally though the orography is only slightly complex. In terms of the relation between meteorological conditions and the forecast error, this leads to an overlap between the overall predictability of the weather situation by the NWP model and local effects on the wind field due to the specific on-site conditions.

As regards local effects, e.g. due to orography causing speed-up effects over hills and channelling effects in valleys, it has to be pointed out that the classification method works with on-site information and is, therefore, unable to tell local flow distortion from global circulation. Hence, local effects are implicitly taken into account and there might be an overlap between the two effects. Among the investigated sites, only one site (Rapshagen) obviously shows this phenomenom and makes it difficult to draw definite conclusions, while the other sites do not show strong signs of local effects in the course of the analysis.

In summary the results for all investigated sites are given in Table 9.3, where the relative rmse for the three weather classes "low passes in the north", "Scandinavian high" and "high centred in north-west" having the most distinct differences in terms of the forecast error are shown. The rather dynamic passage of the low leads to large prediction errors at all sites with the relative rmse ranging from 0.31 to 0.51. Compared with this the two high-pressure situations are related to significantly smaller forecast errors. The high over Scandinavia causes an average rmse ranging from 0.23 to 0.29 with the exception of the unusual error value of 0.43 at Rapshagen presumably related to orographic effects. A high-pressure area centred north-west or west of the sites has also rather small relative rmse of between 0.22 and 0.36. To express the difference between the weather classes in terms of the forecast error, the ratio between the maximum rmse of the low-pressure situation and minimum rmse of the high-pressure is used. It ranges from 1.5 to 1.7, where again Rapshagen is not considered. Hence, the differences are rather profound and the weather type appears to be an important criterion to distinguish different error regimes.

Table 9.3. Overview of the relation between the average forecast error and the meteorological situation for the investigated sites using complete linkage for Fehmarn, Hilkenbrook and Syke and Ward's linkage for Rapshagen[a]

	Number of clusters	Relative rmse "Low passes north of site"	"Scandinavian high"	"High centred in north-west"	rmse ratio max./min
Fehmarn	6	0.39	0.23	0.29	1.7
Hilkenbrook	7	0.51	0.29	0.36	1.8
Rapshagen	7	0.31	0.43*	0.22	2.0*
Syke	7	0.36	0.29	0.24	1.5

[a]The number of clusters is determined by inspection of cluster distances. The relative rmse of those weather classes are shown that typically have high ("low passes in the north") or low ("Scandinavian high", "High-pressure north-west") forecast errors. At Rapshagen the flow situation corresponding to high pressure with easterly winds seems to be influenced by local effects and produces a very unusual, large forecast error marked by ∗.

9.4 Conclusion

In this chapter the quantitative relation between the actual weather situation and the error of the corresponding wind speed prediction has been investigated with methods from synoptic climatology using local measurements of meteorological variables. The main result is that in this framework significant differences in the forecast error for distinct weather situations can be observed. In particular, the expectation can be confirmed that dynamic low-pressure situations with fronts are related to considerably larger prediction errors than weather types that are mainly influenced by rather stationary high-pressure systems. The ratio between the maximum and the minimum error in terms of the average rmse for these situations is between 1.5 and 1.7, which is quite profound.

A classification scheme based on principal component analysis and cluster analysis is successfully applied to automatically divide meteorological situations into different classes. Although the classification procedure uses local information of wind speed, wind direction and atmospheric pressure, the derived classes can be associated with the overall weather situation in most cases by comparing typical days from the clusters with large-scale weather maps. It is important to include information about changing weather conditions by using several measurements of the variables per day. The classification results for the different sites that have been investigated are very consistent.

However, care has to be taken if local effects at the site have a strong influence on the flow, e.g. due to orography causing speed-up effects over hills and channelling effects in valleys. As the classification method works with on-site information, it is unable to tell local flow distortion from global circulation. Hence, local effects are implicitly taken into account and the results may reflect an overlap between the two effects. In this investigation the classification procedure seems to be robust enough to deal with the degree of local inhomogeneity that occurs for sites in northern Germany, including an island site. To exclude local effects it is advisable to prefer wind data measured at 30-m height or higher to 10-m data. For further use of this type of classification scheme it is necessary to systematically evaluate the performance for sites in more complex terrain.

Moreover, it is important to note that this investigation only confirms that there is a relation between the prevailing weather situation and the forecast error of the wind speed for historical data of 1 year. So far, only the 00 UTC prediction run with the lead times 6, 12, 18 and 24 h has been used to assess the daily forecast error because of the limited availability of high-quality data in all variables at all times. Hence, it is desirable for future investigations to shed some light on the behaviour of the prediction horizons beyond 24 h and to confirm the results found here with data from longer periods of time.

With regard to practical applications the advantage of this method is clearly that it works on a rather small set of standard meteorological variables, so that online measurements can be obtained quite easily and cost effectively. This is, together with the fact that the classification scheme is automatic rather than manual, a major prerequisite for a possible operational use of the classification scheme.

So far, measurements of the meteorological variables have been used to determine the weather class. In order to exploit these findings for an estimation of the uncertainty of a wind power prediction, basically two steps are necessary. First of all the predictability of the weather classes themselves has to be evaluated. For prediction purposes it is required to determine the weather type in advance in terms of the predicted wind speed, wind direction and atmospheric pressure. As the forecast quality of pressure and wind direction (e.g. shown by Mönnich [77]) is considerably better than that of wind speed, the prospects of accurately predicting the meteorological class are quite good. This step is non-trivial because the uncertainty of the wind speed prediction which should be provided depending on the weather class is also involved in predicting this weather class. Hence, it must be shown that there is still some advantage in doing so. In a second step the uncertainty of the wind speed forecast has to be transferred to the power forecast. The simplest approach in this direction is to consider the error propagation as applied in Sect. 8.5, where it was shown that the power uncertainty can be estimated rather well by the product of wind speed uncertainty and the derivative of the power curve. The innovative step in terms of the results of this chapter would then be to replace the constant wind speed uncertainties with by weather-type-dependent ones.

10

Smoothing Effects in Regional Power Prediction

Abstract. The investigation in this chapter focusses on the statistical analysis of the power prediction error of an ensemble of wind farms compared with single sites. Due to spatial smoothing effects the relative prediction error decreases considerably. Measurements of the power output of 30 wind farms in Germany show that, as expected, this reduction depends on the size of the region. To generalise these findings an analytical model based on the spatial correlation function of the prediction error is derived to describe the statistical characteristics of arbitrary configurations of wind farms. This analysis reveals that the magnitude of the error reduction depends only weakly on the number of sites and is mainly determined by the size of the region; e.g. for the size of a typical large utility (approximately 730 km in diameter), less than 100 sites are sufficient to have an error reduction of about 50%.

10.1 Introduction

In the previous chapter the typical forecast errors in terms of wind speed and power output are evaluated for single sites. However, in practical use, transmission system operators or electricity traders are mainly interested in the total amount of wind energy to be expected in their supply area for the days to come. Therefore, a prediction of the combined power output of many wind farms distributed over a large region is required. By integrating over an extended area, the errors and fluctuations underlying the measurement and the forecast at single sites cancel out partly. These statistical smoothing effects lead to a reduced prediction error for a region compared with a local forecast. In this chapter the smoothing effects are investigated in greater detail. The analysis of measured data clearly shows a reduction of the error but is constrained to a fixed ensemble of sites. This restriction is overcome by using model ensembles of ficticious sites together with an analytical description of the spatial correlation of the forecast error. Thus, the main parameters that determine the magnitude of the error reduction, namely the size of the region and the number of sites it contains, can easily be varied. The aim is to quantitatively describe the error reduction by spatial smoothing effects for typical wind farm configurations and region sizes. For this purpose the error reduction will mainly be described by the ratio between the forecast error of the regional prediction and the typical forecast error of a single site.

10.2 Ensembles of Measurement Sites

The first approach is to investigate the spatial smoothing effects using measured data from an ensemble of 30 wind farms in the northern part of Germany. The sites are divided into regions of two different types according to typical areas covered by a medium and a large utility. The smaller regions with a diameter of approximately 140 km (see Fig. 10.1) contain three to five measurement sites each. The larger regions are about 350 km in diameter with five to seven sites each. For comparison a very large region containing all sites which has a diameter of about 730 km is formed.

The predicted and measured power output of a region are calculated by summing up the time series for every wind farm located in the region and dividing them by the

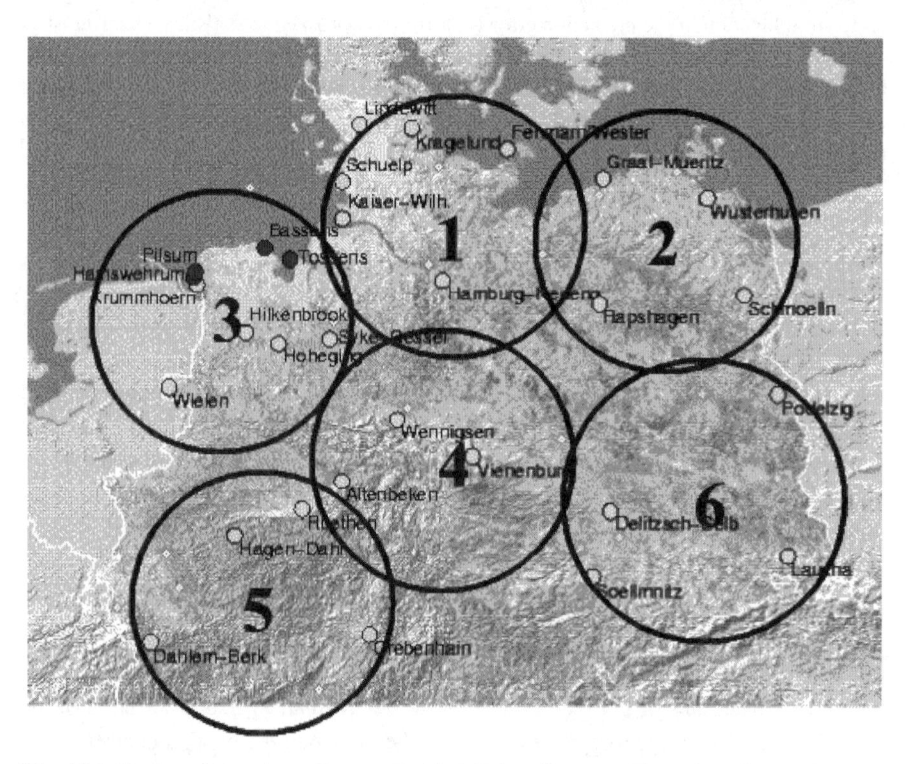

Fig. 10.1. Regions in northern Germany with 140-km diameter. The points denote the measurement sites

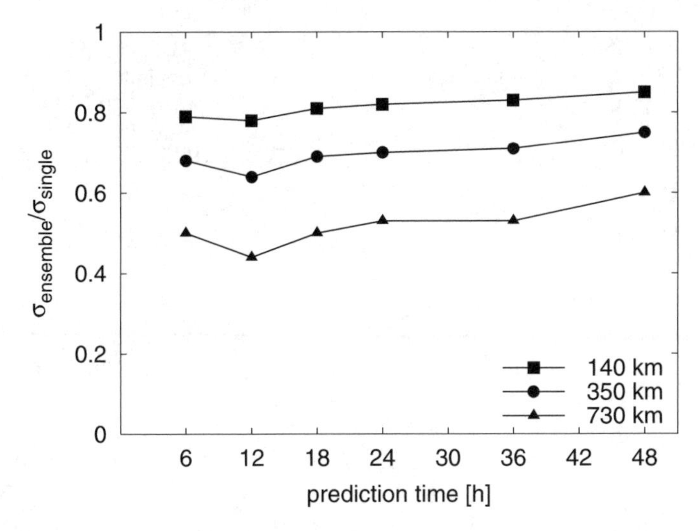

Fig. 10.2. Ratio between standard deviations of error for ensemble and single prediction ($\sigma_{\mathrm{ensemble}}/\sigma_{\mathrm{single}}$) for various region sizes and forecast horizons for the ensemble of measurement sites. $\sigma_{\mathrm{ensemble}}/\sigma_{\mathrm{single}}$ decreases with increasing region size. In all cases the reduction in the regional prediction error is less pronounced for larger prediction times

number of wind farms. Similar to the verification analysis of single sites the standard deviation, $\sigma_{\mathrm{ensemble}}$, of the difference between these two ensemble time series, i.e. the sde as defined in (6.1) and (6.4), gives the regional prediction error. Figure 10.2 shows the results for the different region sizes and various prediction times. The standard deviation of the ensemble, $\sigma_{\mathrm{ensemble}}$, i.e. the regional prediction error, is normalised to the mean standard deviation of the single sites, σ_{single}, according to Chap. 6 and averaged over all regions of the same size. For the given ensemble this ratio decreases with increasing region size, e.g. the 6-h prediction gives an average ratio of 0.79 for the 140-km region, 0.68 for the 350-km region and 0.50 for the 730-km region (Fig. 10.2). In all cases the reduction of the regional prediction error is less pronounced for larger prediction times.

10.3 Model Ensembles

The analysis for a specific set of measurement sites shows a significant decrease in the prediction error compared with a single site. In order to draw general conclusions about other configurations of wind farms, random ensembles of fictitious sites are used. Thus, the size of the regions and the number of wind farms can be varied over large ranges to see how the reduction of the error depends on these parameters. For this purpose it is necessary to establish a proper statistical description of the regional prediction error.

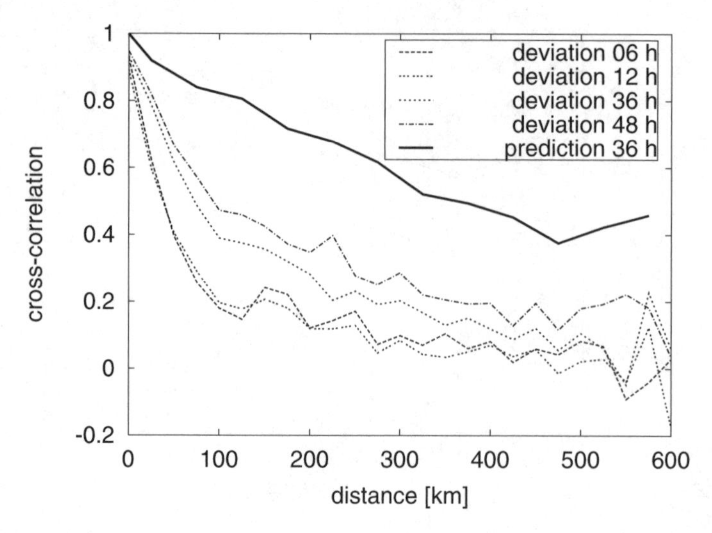

Fig. 10.3. Spatial cross-correlation of prediction error for various prediction times. The cross-correlation coefficients have been averaged over 25 km bins. The values do not go through 1 for $d = 0$ because the distances of wind turbines belonging to the same wind farm are not resolved. For comparison the cross-correlation function of the prediction alone is also shown which decays significantly slower than the error

The key elements connecting the spatial distribution of sites with the regional prediction error are the cross-correlation coefficients r_{xy} of the time series of the pointwise error, i.e. $\epsilon_{\mathrm{P}}(t) := P_{\mathrm{pred}}(t) - P_{\mathrm{meas}}(t)$, between pairs of single sites. If r_{xy} are known for all pairs of sites, the regional forecast error, $\sigma_{\mathrm{ensemble}}$, can easily be calculated using σ of the individual sites by

$$\sigma^2_{\mathrm{ensemble}} = \frac{1}{N^2} \sum_x \sum_y \sigma_x \sigma_y r_{xy} \,, \tag{10.1}$$

where N is the number of sites in the region and σ_x is the standard deviation of the single sites.

To obtain an analytic function describing the dependence of r_{xy} on the site distance, the following procedure is applied. For each pair of the 30 wind farms, the cross-correlation coefficient between pointwise prediction errors is calculated and ordered according to the distance between the two sites x and y. Figure 10.3 shows cross-correlation coefficients versus distance for various forecast horizons, where the data points have been averaged over 25-km bins. For small prediction times (6 and 12 h) the cross-correlation decreases rather rapidly within 150 km, while for longer times (36 and 48 h) the decrease is much slower. This might be due to the growing systematic errors for increasing forecast horizon which give rise to higher spatial correlations.

Note that the spatial cross-correlation coefficients of the measured power alone decay considerably slower with increasing distance than the cross-correlation of the deviations. As shown by Beyer et al. [8], the cross-correlation of the power output of wind turbines is about 0.7 at 130 km, while at the same distance r_{xy} drops to approximately. 0.2 for the 6-h prediction (Fig. 10.3). Hence, due to the large-scale weather patterns which simultaneously affect many wind farms the spatial correlation of the power output is considerably higher than that of the power prediction error.

The cross-correlation of the regional prediction error is approximated by fitting analytic functions of the form $r_{xy} = a \cdot e^{-d/b}$ (a and b are fitparameters and d is the distance between the two sites) to the cross-correlation coefficients derived from the measured data. It turns out that piecewise exponentials lead to a suitable fit to the data points. Now the correlation function r_{xy} based on the fitted data together with (10.1) allows for calculating the prediction error, σ_{ensemble}, of the model regions with fictitious wind farms. The individual forecast errors σ_x of the fictitious wind farms are assumed to be equal, which means that they all have the same weight. The geographical coordinates of the model ensembles are chosen randomly. Each result given in the following represents an average value over 10 realisations of ensembles with fixed size and number of sites.

Figure 10.4 shows the ratio between the regional error and the mean of single sites, $\sigma_{\text{ensemble}}/\sigma_{\text{single}}$, for two regions with different sizes versus the number of sites in the region. The cross-correlation function r_{xy} based on the 36-h forecast was used. Obviously, $\sigma_{\text{ensemble}}/\sigma_{\text{single}}$ approaches a saturation level for increasing

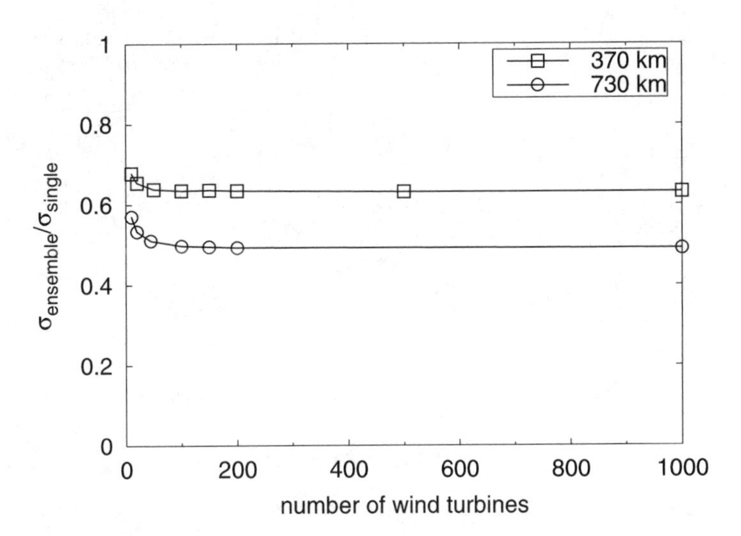

Fig. 10.4. Ratio $\sigma_{\text{ensemble}}/\sigma_{\text{single}}$ versus number of sites for the model ensembles. Each data point represents an average over 10 ensembles. The fitted cross-correlation function for the 36-h forecast was used. A saturation level is reached for a relatively small number of sites

number of wind farms. This limit is already reached for a rather small number of wind farms. Beyond that point the error reduction does practically not depend on the number of sites, e.g. for the size of a typical large utility (approximately. 730 km) about 100 sites are sufficient to have a constant level of 0.49.

For uncorrelated sources, one would expect $\sigma_{\text{ensemble}}/\sigma_{\text{single}} = N^{-\frac{1}{2}}$ showing no saturation for large N. Let the limit value for a model region be σ_{sat}; then the number of uncorrelated sources with at least the same signal-to-noise ratio is simply given by

$$N = \text{int}\left(\sigma_{sat}^{-2}\right) + 1 ,$$

where the operator int truncates the digits. As illustrated in Fig. 10.5, the limit value of the largest region, which is 730 km in diameter, corresponds to about 5 (exactly 4.15) uncorrelated sources, and that of one of the smaller regions to 3 (2.51). Thus, due to the finite correlations of the prediction error between distant sites the regional error reduction is far less pronounced than for uncorrelated sites.

The experimental data in Sect. 10.2 agree well with these calculations. Comparing the 36-h data in Fig. 10.2 with the values corresponding to the same number of sites, i.e. 30, in Fig. 10.4 gives $\sigma_{\text{ensemble}}/\sigma_{\text{single}} = 0.71$ for the 360-km region for both ensembles. For the 730-km region this ratio is 0.53 for the ensemble of measurement sites and 0.52 for the model ensembles.

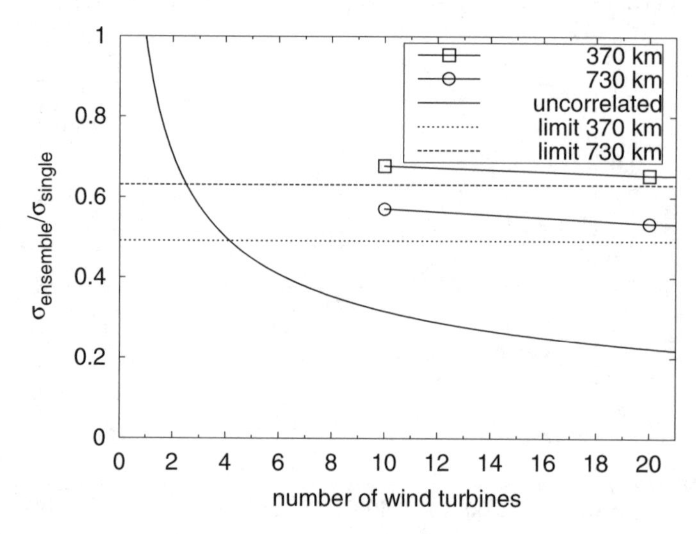

Fig. 10.5. Ratio $\sigma_{\text{ensemble}}/\sigma_{\text{single}}$ versus number of sites for the model ensembles with the $N^{-\frac{1}{2}}$ behaviour of uncorrelated sites. The horizontal lines indicate the limit values of the model regions. The intersection of these lines with $N^{-\frac{1}{2}}$ provides the number of uncorrelated sources that would lead to the same error reduction

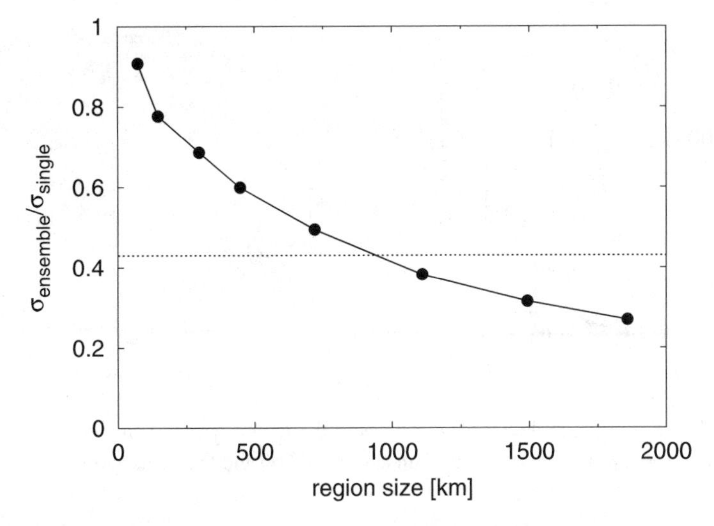

Fig. 10.6. Saturation values of $\sigma_{\text{ensemble}}/\sigma_{\text{single}}$ (4000 sites) for the 36-h forecast. Each data point represents an average over 10 ensembles. The dashed line illustrates the ratio $\sigma_{\text{ensemble}}/\sigma_{\text{single}}$ determined for the ensemble of all wind farms in Germany

As expected, the saturation level decreases with increasing size of the region. This is illustrated in Fig. 10.6, where the limit values for regions with different extensions containing 4000 sites are shown. There is a rapid decay for extensions below 500 km.

The distribution of distances within the ensemble of wind farms plays a crucial role in explaining the saturation level. Figure 10.7 shows typical frequency distributions of the distances in a 360-km region for different numbers of randomly distributed wind farms. The distributions correspond to specific ensembles, i.e. another realisation has a different distribution. Figure 10.7 clearly indicates that the distribution functions converge for an increasing number of wind farms. Beyond a certain number of wind farms in the region, the characteristic distribution of distances approaches a limit distribution so that the regional error given by (10.1) does no longer change. If the region has a certain "population density" of wind farms, it contains enough pairs to represent all possible correlations which contribute to σ_{ensemble}. Therefore, adding more sites to the region does not reduce the regional error, as would be expected for uncorrelated sites, and a saturation is reached.

The limit distribution has a mean value which is typically about half the diameter of the region. As the distribution is concentrated around the mean, most contributions to σ_{ensemble} come from this interval. For increasing region size this leads to a decreasing saturation level due to the exponential decay of the cross-correlation function (Fig. 10.3).

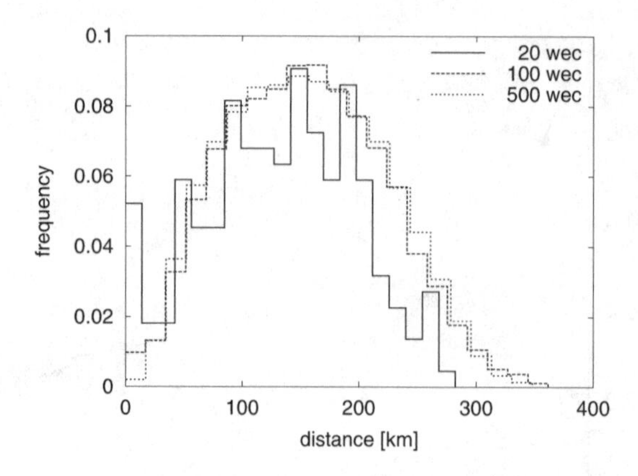

Fig. 10.7. Frequency distribution of distances in the region with a diameter of 360 km. The curves represent randomly chosen coordinates of 20, 100 and 500 wind farms

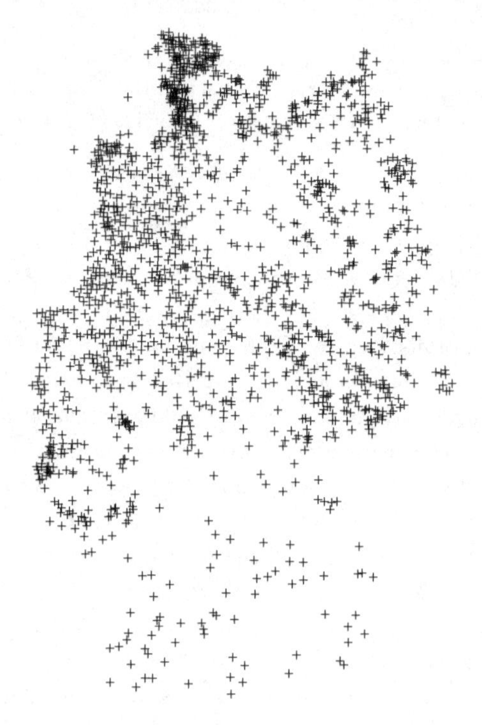

Fig. 10.8. Distribution of wind turbines in Germany in 1999. The crosses denote ZIP-code areas of the site locations

10.3.1 Distribution of German Wind Farms

Finally, the real distribution of wind farms in Germany in the year 1999 is considered as a special model ensemble (Fig. 10.8) and the regional forecast error is compared with that of a single site as above. For the 36-h prediction this gives

$\sigma_{\text{ensemble}}/\sigma_{\text{single}} = 0.43$. This ratio is illustrated by a dashed line in Fig. 10.6, giving a corresponding region size of 930 km which rather exactly matches the north to south extension of Germany. Hence, this ratio agrees well with the theoretically found value.

Note that the imbalance in the distribution of sites in the north and south (Fig. 10.8) does not affect the resulting error reduction factor $\sigma_{\text{ensemble}}/\sigma_{\text{single}}$ as the critical number of around 100 for a region of that size is, of course, exceeded.

10.4 Conclusion

The statistical smoothing effects of the prediction uncertainty that arise if a wind power forecast is made for a region with spatially distributed wind farms lead to a reduction in the error of the aggregated power prediction compared with a single site. For an ensemble of wind farms where the analysis is based on measured data, the improvement of the prediction is noticeable even for small regions and only a few sites. Using model ensembles with randomly chosen locations permits to generalise the results in order to identify the impact of the two main parameters, namely the spatial extension of the region and the number of sites it contains, together with their distribution. The magnitude of the reduction does strongly depend on the size of the region; i.e. the larger the region, the larger the reduction. As regards the number of sites contained in the area, a saturation level is already reached for a comparatively small number of wind farms. This means that only a few sites are sufficient to determine the magnitude of the improvement in the power prediction.

The results of the analysis show that for regions with a sufficient number of preferably equally distributed wind farms, it is now possible to estimate the regional smoothing effect of the wind power prediction error by just considering the size of the region in question. As the error reduction is due to statistical effects, the power prediction for a region improves on average. However, single events such as an incorrectly predicted storm front crossing the entire area lead to a coherent behaviour of the wind turbines. In this case the correlation of the prediction error between the sites is, of course, higher than on average and spatial smoothing is less pronounced.

This investigation was carried out with measured data for northern Germany and output of the NWP model of the German Weather Service. The regions are characterised by mainly flat terrain with no mountains separating the area in different weather regimes. The parameters, especially the decay length of the spatial cross-correlation of the deviations, are supposed to depend on the typical length scales of the weather patterns crossing the region and on the NWP model which may have special characteristics. In the mid-latitudes, where the weather conditions are driven by high-and low-pressure systems, the general patterns of circulation types are comparable to our situation. Thus, the general behaviour of the cross-correlation function between the sites which determines the degree of statistical smoothing and the saturation values should be universal so that the qualitative results found in this investigation are applicable to other areas in the mid-latitudes as well.

11
Outlook

The field of wind power prediction is developing quite dynamical, so that the state of the art is continously moving forward and each documentation about the current status is merely a snapshot. The challenges of the future require further research and development as wind energy is claiming a growing share in the electricity supply of many countries. With regard to wind power predictions, the fields of future work are already on the agenda covering a number of interesting topics such as upscaling of representative sites, offshore application, uncertainty estimation and meso-scale modelling in complex terrain. Each of these topics deserves some further explanation.

Wind power predictions are usually applied to obtain forecasts of the aggregated power production of a number of wind farms in an area, e.g. by a transmission system operator who is interested in keeping the grid balanced. Calculating a regional power prediction based on each wind farm contained in the area would be a straightforward way, but it has some technical limitations, e.g. regarding the availability of measurement data or the finite computer ressources. Consequently, it has become a well-established strategy to select a number of representative sites where predictions are made and then scale these predictions up to the anticipated regional production. The challenge here is, of course, to carefully select the representative sites and find the optimal weight for each of them. As already pointed out in Chap. 10 this requires a firm statistical foundation in the superposition of dependent time series and a quantitative description of the error reduction by spatial smoothing effects. The properties of the region have to be covered by taking the climatology of the wind regime into account. Hence, developping advanced upscaling algorithms requires an understanding of the temporal and spatial structure of wind fields in order to estimate the variations of wind power over an area. In addition, the technical details, e.g. coordinates, hub height and installed power, of the wind turbines in the region have to be available and embedded properly into the upscaling procedure.

Going offshore is a challenge in virtually every respect. The installed power of the planned wind farms at sea will be of the order of magnitude of that of nuclear power plants. So there is no doubt that accurate forecasts of their power production are beneficial. As the meteorological conditions of the marine boundary layer are only partly understood, the concepts developed for onshore wind energy applications, as discussed in Chaps. 3 and 7, have to be modified. In particular, the variable surface roughness of the ocean and the large heat capacity of the water have a significant impact on the vertical wind speed profile. Thus, with regard to wind power predictions, new models to describe the vertical profile are required which consider the characteristic transfer of momentum and heat between sea surface and atmosphere. First investigations show that the accuracy of the wind speed forecast is rather promising. However, in contrast to the onshore case, wind power predictions offshore will typically be made for single wind farms and, thus, cannot take advantage of spatial smoothing effects. Therefore, improving the prediction quality offshore is an important issue. In addition, the considerable spatial dimensions of offshore wind farms might strongly influence the wind field within the wind farms and in their vicinity. Hence, the presence of the wind farm is expected to have a macroscopic impact, e.g. due to wake effects or the formation of internal boundary layers. These phenomena have to be investigated and included into wind power prediction systems.

The uncertainty of wind power predictions is a major issue in this book (Chaps. 6, 8 and 9) and also for the future. Analysing the uncertainty of a wind power prediction provides considerable insight into the basic properties of the prediction system, such as the effect of the non-linear power curve. Moreover, if the uncertainty information is tailor-made for the individual forecast situation, it is a useful additional information for the end user who can then better estimate the risk of trusting in the prediction. The uncertainty of the power prediction is the cumulated effect of the errors introduced by input variables (e.g. wind speed) that are propagated through the prediction system and of errors in the model parameters (e.g. certified power curve). In order to quantitatively describe the resulting error both error sources have to be kept apart. It has already become clear that the uncertainty of the wind speed prediction plays an important role in the final power prediction error. Therefore, approaches to assess the wind speed uncertainty in terms of the prevailing forecast situation are promising. The question is, of course, what exactly constitutes the forecast situation? In order to find a feasible answer a wide range of approaches are being tested involving weather classification schemes, methods from synoptic climatology and ensemble predictions. The latter approach is a rather elegant way to include the inherent dynamic properties of the numerical weather prediction (NWP) system, which are in fact partly responsible for the prediction errors, into the assessment of the forecast uncertainty. However, further research and development will reveal the benefits of either of these methods.

For certain wind farm sites the rather coarse numerical grid of the operational NWP systems does not sufficiently cover the local wind conditions. This can be due to the complexity of the terrain with orographic structures that significantly change the wind field on a rather small scale. If on top of that the solar radiation is high, the traditional NWP systems have difficulties to accurately describe the situation. One important example is the situation in Spain where quite a number of wind farms are erected in areas that are very challenging to predict. For these applications the operational use of meso-scale atmospheric models appears reasonable. The resolution of these models can be decreased to a few hundred metres, but what is more important is that they are designed to include small-scale effects such as thermally driven convection in valleys. As mentioned in Chap. 2, meso-scale models are readily available but they have to be adapted and implemented into wind power prediction systems. Of course, the effort to run meso-scale models in an operational setup is high not only due to computational costs but also due to the large amount of input data required, e.g. digitized surface maps. However, this effort might be small compared with the benefits from an accurate prediction.

In conclusion, the field of wind power prediction will develop along the lines defined by the challenges that are brought up by the large-scale utilisation of wind energy in different areas of the world. It is up to both researchers and industry to exploit the considerable potential of wind power predictions and take advantage of the information about the energy of tomorrow.

A

Definition of Statistical Quantities

A.1 General Statistical Quantities

Let $\{x_i\}$, with $i = 1, \ldots, N$, be a given discrete time series of length N. $\{x_i\}$ is regarded as a finite sample from a continuous, infinite and stationary time series $x(t)$.

Mean

The true mean μ of $x(t)$ is estimated by the mean value $\overline{x}$ of $\{x_i\}$, which is defined by

$$\overline{x} = \frac{1}{N} \sum_{i=1}^{N} x_i \, . \tag{A.1}$$

Standard Deviation

The standard deviation $\sigma(x)$ of $x(t)$ is estimated by the standard deviation σ_x of the sample; hence,

$$\sigma_x = \sqrt{\frac{1}{N-1} \sum_{i=1}^{N} (x_i - \overline{x})^2} \, , \tag{A.2}$$

where $\overline{x}$ denotes the mean according to (A.1). σ_x^2 will be referred to as variance.

Cross-correlation Coefficient

The cross-correlation measures the degree of linear dependence between two times series $x(t)$ and $y(t)$. It is defined by

$$r_{xy} = \frac{1}{\sigma_x \, \sigma_y} \frac{1}{N} \sum_{i=1}^{N} (x_i - \overline{x}) \, (y_i - \overline{y}) \; . \tag{A.3}$$

A.2 Error Measures

Error measures describe the average behaviour of deviations between predicted and measured values. The time series to be analysed is the difference between prediction, x_{pred}, and measurement, x_{meas}, i.e. the pointwise error at time i defined by

$$\epsilon_i = x_{\text{pred,i}} - x_{\text{meas,i}} \; . \tag{A.4}$$

Root Mean Square Error (rmse)

The root mean square error evaluates the squared difference between x_{pred} and x_{meas}; hence,

$$\text{rmse} = \sqrt{\frac{1}{N} \sum_{i=1}^{N} \left(x_{\text{pred,i}} - x_{\text{meas,i}}\right)^2} = \sqrt{\overline{\epsilon^2}} \; . \tag{A.5}$$

Bias

The bias is the difference between the mean values of x_{pred} and x_{meas}, i.e.

$$\text{bias} = \overline{x_{\text{pred}}} - \overline{x_{\text{meas}}} = \overline{\epsilon} \; . \tag{A.6}$$

Standard Deviation of Error (sde) and its Decomposition

The standard deviation of error is given by

$$\text{sde} = \sigma(\epsilon) = \sqrt{\overline{\epsilon^2} - \overline{\epsilon}^2} \; . \tag{A.7}$$

By using simple algebra the variance of error, i.e. sde^2, can be split into two parts, namely the sdbias defined by the difference between the standard deviations of x_{pred} and x_{meas} and the dispersion, disp:

$$
\begin{aligned}
\mathrm{sde}^2 &= \sigma^2(\epsilon) \\
&= \overline{(\epsilon - \overline{\epsilon})^2} \\
&= \overline{\left(x_{\mathrm{pred}} - \overline{x_{\mathrm{pred}}} - (x_{\mathrm{meas}} - \overline{x_{\mathrm{meas}}})\right)^2} \\
&= \overline{(x_{\mathrm{pred}} - \overline{x_{\mathrm{pred}}})^2} + \overline{(x_{\mathrm{meas}} - \overline{x_{\mathrm{meas}}})^2} - 2\overline{(x_{\mathrm{pred}} - \overline{x_{\mathrm{pred}}})\,(x_{\mathrm{meas}} - \overline{x_{\mathrm{meas}}})} \\
&= \sigma^2(x_{\mathrm{pred}}) + \sigma^2(x_{\mathrm{meas}}) - 2\sigma(x_{\mathrm{pred}})\,\sigma(x_{\mathrm{meas}})\,r(x_{\mathrm{pred}}, x_{\mathrm{meas}}) \\
&= \left(\sigma^2(x_{\mathrm{pred}}) + \sigma^2(x_{\mathrm{meas}}) - 2\sigma(x_{\mathrm{pred}})\,\sigma(x_{\mathrm{meas}})\right) \\
&\qquad + 2\sigma(x_{\mathrm{pred}})\,\sigma(x_{\mathrm{meas}}) - 2\sigma(x_{\mathrm{pred}})\,\sigma(x_{\mathrm{meas}})\,r(x_{\mathrm{pred}}, x_{\mathrm{meas}}) \\
&= \underbrace{\left(\sigma(x_{\mathrm{pred}}) - \sigma(x_{\mathrm{meas}})\right)^2}_{\mathrm{sdbias}^2} + \underbrace{2\sigma(x_{\mathrm{pred}})\,\sigma(x_{\mathrm{meas}})(1 - r(x_{\mathrm{pred}}, x_{\mathrm{meas}}))}_{\mathrm{disp}^2}
\end{aligned}
\tag{A.8}
$$

Using (A.5), (A.6), (A.7) and (A.8), the rmse can be written as

$$
\mathrm{rmse} = \sqrt{\mathrm{bias}^2 + \mathrm{sdbias}^2 + \mathrm{disp}^2}\ . \tag{A.9}
$$

B

Statistical Testing

The type of probability distribution of the variables plays an important role in the interpretation of the results, in particular in connection with confidence intervals. In some cases an assumption concerning the distribution of the variable in question can be made, e.g. the deviations between prediction and measurement are commonly expected to be normally distributed. Hypotheses of this kind can be tested with statistical methods. The idea is to formulate a hypothesis H_0, such as "$H_0 =$ A given sample $\{x_i\}$ is drawn from a distribution of type Y". Then a confidence level $1 - \alpha$ is set, where α is the probability of rejecting the hypothesis although it is true. The outcome of the test is either that H_0 is rejected or it is not rejected at the given confidence level. So the crucial point is that a statistical test cannot prove the hypothesis; i.e. even if the hypothesis is not falsified by the test, it can still be wrong. The value of these tests is to add some more confidence to the statement of the hypothesis.

B.1 The χ^2 Test

The χ^2 test is a popular form of hypothesis testing and is described in standard books on statistics, e.g. [121]. The test is used in this work as a so-called goodness-of-fit test to check whether a sample could have been drawn from a normal distribution. Hence, the null hypothesis is "$H_0 =$ The given random sample $\{x_i\}$ is drawn from a normal distribution with mean $\overline{x}$ and variance σ^2". The basic idea of the χ^2 test is to compare observed frequencies, h_j, with the theoretically expected probabilities, p_j, and provide criteria to decide if they show significant differences.

First of all, the N sample values $\{x_i\}$ are divided into k bins of equal width Δe. Let e_j be the upper boundary of bin j; then the theoretical probability, assuming that the distribution was Gaussian, is given by $p_j = F(e_j) - F(e_{j-1})$, where $F(e_j) = \Phi\left((e_j - \overline{x})/\sigma\right)$ is the cumulated distribution function of a Gaussian distribution. Then the random variable

$$\chi_s^2 = \sum_j^k \frac{(h_j - N\,p_j)^2}{N\,p_j} \tag{B.1}$$

is defined which compares the empirically found frequencies, h_j, with the expected Gaussian probabilities, p_j. The variable χ_s^2 is approximately χ^2 distributed with $k-1$ degrees of freedom. A prerequisite for this approximation is $Np_j \geq 5$. The χ^2 distribution is defined as the distribution of the sum of the squares of independent standard normal variables and its values are usually tabulated, e.g. in [121].

The probability of wrong rejection α is selected. It implicitly defines the interval in which realisations of χ_s^2 according to (B.1) are rejected. Hence, the condition

$$F_{\chi^2}(\chi^2 \geq \chi^2_{\alpha,k-1}) = \alpha\,, \tag{B.2}$$

where F_{χ^2} is the cumulated χ^2 distribution, gives a critical point, $\chi^2_{\alpha,k-1}$, and the probability to find values beyond this point is α. Typically, α is set to 0.10, 0.05 or 0.01 and is called the significance level. The corresponding values of $\chi^2_{\alpha,k-1}$ are tabulated.

After $\chi^2_{\alpha,k-1}$ is determined, the last step is to check whether the realisation of χ_s^2 is smaller than this critical point; i.e. if

$$\chi_s^2 \leq \chi^2_{\alpha,k-1}\,, \tag{B.3}$$

there is no objection to the assumption that the sample stems from a Gaussian distribution.

B.2 The Lilliefors Test

The Lilliefors test is also a goodness-of-fit test used to test a given distribution for normality. The test compares the empirical cumulative distribution of the sample $\{x_i\}$ with a normal cumulative distribution having the same mean and variance as $\{x_i\}$. Compared to the χ^2-test it has the advantage that the sample size is allowed to be relatively small.

The test statistics of the Lilliefors test is defined as

$$D = \max_j |P_j - H_j| \tag{B.4}$$

where $P_j = \Phi\left((x_j - \overline{x})/\sigma\right)$ is the cumulative distribution of the normal standard distribution and H_j the empirical frequency of events with $x \leq x_j$. D according to (B.4) is tested against a critical value D_α at the desired confidence level α obtained from a table. If $D > D_\alpha$ the hypothesis that the distribution is normal will be rejected.

The known limitations of the Lilliefors test (which is similar to a test called Kolmogorov-Smirnov test) is that it tends to be more sensitive near the centre of the distribution than at the tails.

B.3 The F Test

In order to test whether several sample means show statistically significant differences the so-called F test is used (described, e.g., in [121]). The idea of this test is to compare the within variance among the members of one sample to the in-between variance among the means of the different samples. The variance between the sample means is "explained" by the fact that the samples might come from different populations, while the variance within one sample is "unexplained" in the sense that it is due to random fluctuations around the mean. Hence, the F test is based on evaluating the ratio

$$F = \frac{\text{explained variance}}{\text{unexplained variance}} . \tag{B.5}$$

Typically, the hypothesis to be tested is formulated as "H_0 = The sample means belong to the same distribution."

To compare r different means derived from samples of unequal size the following notation is used: let $\{x_i^j\}$, with $i = 1, \ldots, N_j$ and $j = 1, \ldots, r$, denote the jth sample, e.g. in Sect. 9.3 the daily error values of one cluster; then $\overline{x}^j$ is the sample mean.

The average variance of the different samples from their respective mean values is called pooled variance and is defined by

$$\sigma_p^2 = \frac{\sum_{j=1}^{r} \sum_{i=1}^{N_j} (x_i^j - \overline{x}^j)^2}{\sum_{j=1}^{r} (N_j - 1)} . \tag{B.6}$$

The variance between the sample means is defined by

$$\sigma_m^2 = \frac{\sum_{j=1}^{r} N_j (\overline{x}^j - \overline{\overline{x}})^2}{r - 1} , \tag{B.7}$$

where $\overline{\overline{x}}$ denotes the mean of all x_i^j, i.e.

$$\overline{\overline{x}} = \frac{\sum_{j=1}^{r} N_j \overline{x}^j}{\sum_{j=1}^{r} N_j} . \tag{B.8}$$

Now the average within-variance of the individual samples is compared to the variance of the sample means defining the F ratio in (B.5) by

$$F = \frac{\sigma_m^2}{\sigma_p^2} . \tag{B.9}$$

The F ratio is determined for the given samples. If H_0 is true and the sample means are the same, then F is around 1. If the mean values are from different distributions and, hence, H_0 is not true, the variance among the mean values, σ_m^2, will be larger compared with the within-sample variance, σ_p^2, and F in (B.9) is expected to be greater than 1. The F distribution determines how large F is allowed to be under the assumption that H_0 is true. Hence, for a pre-defined significance level α, i.e. the probability that H_0 is rejected though it is true, the critical value F_α can be inferred from a tabulated F distribution with the suitable degree of freedom. If the condition

$$F \leq F_\alpha \tag{B.10}$$

holds, H_0 is not rejected.

Note that the hypothesis H_0 is "the means are equal" and, thus, if H_0 is rejected the alternative hypothesis H_1: "the means are NOT equal" is confirmed. However, this is only the first step because this test does not tell which pairs of mean values are different. This has to be tested by constructing simultaneous confidence intervals around each of the mean values. Using *Scheffe's multiple comparisons* [121] the differences between all pairs of sample means can be evaluated, and with 95% confidence the following statement is true for all sample pairs (k, j) simultaneously:

$$\mu_k - \mu_j = (\overline{x}^k - \overline{x}^j) \pm \sqrt{(r-1)\,F_{0.05}}\,\sigma_p \sqrt{\frac{1}{N_k} + \frac{1}{N_j}}\,, \tag{B.11}$$

where $\mu_k - \mu_j$ is the difference between the true means of the two underlying populations, $\overline{x}^k - \overline{x}^j$ is the difference between the sample means, r is the number of different samples, $F_{0.05}$ is the value of the F distribution at the significance level 0.05, σ_p^2 is the pooled variance (B.6) and N_k,N_j are the sample sizes.

Thus, if the difference $\overline{x}^k - \overline{x}^j$ exceeds the confidence range given by (B.11), the two samples $\{x_i^k\}$ and $\{x_i^j\}$ are believed to originate from two different distributions and the mean values $\overline{x}^k$ and $\overline{x}^j$ are regarded as significantly different.

B.4 *F* Test Results from Sect. 9.3

Table B.1. *F* ratios (B.9) of investigated stations (see Sect. 9.3)

	Number of clusters	Linkage type	*F* ratio
Fehmarn	6	Complete	7.73
	6	Ward's	5.65
Hilkenbrook	7	Complete	6.23

Table B.2. Results of Scheffe's multiple comparison for daily rmse for Fehmarn clustering with complete linkage[a]

Cluster$_i$	Cluster$_j$	Difference of means$_{ij}$	Significance	Lower boundary	Upper boundary
1	2	−0.0709	1.000	−0.7788	0.6371
	3	0.4120	0.063	−0.0119	0.8360
	4	0.7619(*)	0.001	0.2061	1.3176
	5	−0.0875	0.999	−0.7433	0.5682
	6	0.4930(*)	0.011	0.0690	0.9169
2	1	0.0709	1.000	−0.6371	0.7788
	3	0.4829	0.297	−0.1702	1.1360
	4	0.8327(*)	0.017	0.0873	1.5782
	5	−0.0167	1.000	−0.8394	0.8060
	6	0.5638	0.141	−0.0893	1.2170
3	1	−0.4120	0.063	−0.8360	0.0119
	2	−0.4829	0.297	−1.1360	0.1702
	4	0.3498	0.323	−0.1341	0.8338
	5	−0.4996	0.167	−1.0957	0.0965
	6	0.0809	0.983	−0.2433	0.4051
4	1	−0.7619(*)	0.001	−1.3176	−0.2061
	2	−0.8327(*)	0.017	−1.5782	−0.0873
	3	−0.3498	0.323	−0.8338	0.1341
	5	−0.8494(*)	0.006	−1.5455	−0.1534
	6	−0.2689	0.630	−0.7529	0.2151
5	1	0.0875	0.999	−0.5682	0.7433
	2	0.0167	1.000	−0.8060	0.8394
	3	0.4996	0.167	−0.0965	1.0957
	4	0.8494(*)	0.006	0.1534	1.5455
	6	0.5805	0.062	−0.0156	1.1766
6	1	−0.4930(*)	0.011	−0.9169	−0.0690
	2	−0.5638	0.141	−1.2170	0.0893
	3	−0.0809	0.983	−0.4051	0.2433
	4	0.2689	0.630	−0.2151	0.7529
	5	−0.5805	0.062	−1.1766	0.0156

[a]The standard software SPSS has been used. "*" denotes that the difference between the means is significant at the level $\alpha = 0.05$. Cluster$_i$ and cluster$_j$ refer to the cluster numbers used in Sect. 9.3. If "significance" is smaller than $\alpha = 0.05$, the mean values of the two clusters i and j are regarded as significantly different. Lower boundary and upper boundary refer to the boundaries of the 95% confidence interval given by B.11. This interval does not contain the origin if the clusters are significantly different.

Table B.3. Results of Scheffe's multiple comparison for daily rmse, for Fehmarn with Ward's linkage[a]

Cluster_i	Cluster_j	Difference of means_{ij}	Significance	Lower boundary	Upper boundary
1	2	0.1725	0.959	−0.3920	0.7369
	3	−0.2504	0.757	−0.7669	0.2662
	4	−0.0253	1.000	−0.5898	0.5391
	5	−0.3974	0.592	−1.087	0.2926
	6	−0.4223	0.206	−0.9474	0.1029
2	1	−0.1725	0.959	−0.7369	0.3920
	3	−0.4228(*)	0.048	−0.8439	−0.0017
	4	−0.1978	0.861	−0.6764	0.2809
	5	−0.5699	0.097	−1.191	0.0519
	6	−0.5947(*)	0.001	−1.026	−0.1631
3	1	0.2504	0.757	−0.2662	0.7669
	2	0.4228(*)	0.048	0.0017	0.8439
	4	0.2250	0.670	−0.1961	0.6462
	5	−0.1471	0.982	−0.7258	0.4316
	6	−0.1719	0.782	−0.5387	0.1949
4	1	0.0253	1.000	−0.5391	0.5898
	2	0.1978	0.861	−0.2809	0.6764
	3	−0.2250	0.670	−0.6462	0.1961
	5	−0.3721	0.549	−0.9939	0.2497
	6	−0.3969	0.095	−0.8286	0.0347
5	1	0.3974	0.592	−0.2926	1.0875
	2	0.5699	0.097	−0.0519	1.1917
	3	0.1471	0.982	−0.4316	0.7258
	4	0.3721	0.549	−0.2497	0.9939
	6	−0.0248	1.000	−0.6112	0.5616
6	1	0.4223	0.206	−0.1029	0.9474
	2	0.5947(*)	0.001	0.1631	1.0264
	3	0.1719	0.782	−0.1949	0.5387
	4	0.3969	0.095	−0.0347	0.8286
	5	0.0248	1.000	−0.5616	0.6112

[a]The standard software SPSS has been used. "*" denotes that the difference between the means is significant at the level $\alpha = 0.05$. Cluster_i and cluster_j refer to the cluster numbers used in Sect. 9.3. If "significance" is smaller than $\alpha = 0.05$, the mean values of the two clusters i and j are regarded as significantly different. Lower boundary and upper boundary refer to the boundaries of the 95% confidence interval given by B.11. This interval does not contain the origin if the clusters are significantly different.

Table B.4. Results of Scheffe's multiple comparison for daily rmse for Hilkenbrook with complete linkage[a]

Cluster_i	Cluster_j	Difference of means_{ij}	Significance	Lower boundary	Upper boundary
1	2	−0.2309	0.372	−0.5544	0.0925
	3	−0.1616	0.964	−0.6443	0.3212
	4	−0.2806(*)	0.035	−0.5509	−0.0104
	5	−0.2365	0.852	−0.7561	0.2831
	6	−0.7372(*)	0.000	−1.1838	−0.2906
	7	−0.2088	0.580	−0.5516	0.1341
2	1	0.2309	0.372	−0.0925	0.5544
	3	0.0694	1.000	−0.4218	0.5605
	4	−0.0497	0.999	−0.3347	0.2353
	5	−0.0055	1.000	−0.5330	0.5219
	6	−0.5063(*)	0.017	−0.9619	−0.0506
	7	0.0222	1.000	−0.3324	0.3767
3	1	0.1616	0.964	−0.3212	0.6443
	2	−0.0694	1.000	−0.5605	0.4218
	4	−0.1191	0.990	−0.5769	0.3388
	5	−0.0749	1.000	−0.7126	0.5628
	6	−0.5756	0.053	−1.1553	0.0041
	7	−0.0472	1.000	−0.5513	0.4569
4	1	0.2806(*)	0.035	0.0104	0.5509
	2	0.0497	0.999	−0.2353	0.3347
	3	0.1191	0.990	−0.3388	0.5769
	5	0.0442	1.000	−0.4525	0.5408
	6	−0.4566(*)	0.021	−0.8761	−0.0370
	7	0.0719	0.994	−0.2349	0.3787
5	1	0.2365	0.852	−0.2831	0.7561
	2	0.0055	1.000	−0.5219	0.5330
	3	0.0749	1.000	−0.5628	0.7126
	4	−0.0442	1.000	−0.5408	0.4525
	6	−0.5007	0.203	−1.1115	0.1100
	7	0.0277	1.000	−0.5118	0.5672
6	1	0.7372(*)	0.000	0.2906	1.1838
	2	0.5063(*)	0.017	0.0506	0.9619
	3	0.5756	0.053	−0.0041	1.1553
	4	0.4566(*)	0.021	0.0370	0.8761
	5	0.5007	0.203	−0.1100	1.1115
	7	0.5284(*)	0.014	0.0588	0.9980

(*cont.*)

Table B.4. Contd.

$Cluster_i$	$Cluster_j$	Difference of $means_{ij}$	Significance	Lower boundary	Upper boundary
7	1	0.2088	0.580	−0.1341	0.5516
	2	−0.0222	1.000	−0.3767	0.3324
	3	0.0472	1.000	−0.4569	0.5513
	4	−0.0719	0.994	−0.3787	0.2349
	5	−0.0277	1.000	−0.5672	0.5118
	6	−0.5284(*)	0.014	−0.9980	−0.0588

[a]The standard software SPSS has been used. "*" denotes that the difference between the means is significant at the level $\alpha = 0.05$. $Cluster_i$ and $cluster_j$ refer to the cluster numbers used in Sect. 9.3. If "significance" is smaller than $\alpha = 0.05$, the mean values of the two clusters i and j are regarded as significantly different. Lower boundary and upper boundary refer to the boundaries of the 95% confidence interval given by B.11. This interval does not contain the origin if the clusters are significantly different

References

1. J. Ainslie. Calculating the flow field in the wake of wind turbines. *J. Wind Engg. Ind. Aerodyn.*, **27**:213–224, 1988.
2. S.P. Arya. *Introduction to Micrometeorology*. Academic Press, San Diego, 2001.
3. Bundesverband WindEnergie (German Wind Energy Association). `www.wind-energie.de`.
4. B. Bailey, M.C. Brower, and J. Zack. Short-term wind forecasting—development and application of a mesoscale model. In *Proc. Eur. Wind Energy Conf. EWEC*, p. 1062, Nice, 1999.
5. R.G. Barry and A.H. Perry. *Synoptic Climatology*. Methuen, London, 1973.
6. A.C.M. Beljaars and F.C. Bosveld. Cabouw data for validation of land surface parameterization schemes. *J. Clim.*, **10**:1172–1193, 1997.
7. H.G. Beyer, D. Heinemann, H. Mellinghoff, K. Mönnich, and H.P. Waldl. Forecast of regional power output of wind turbines. In *Proc. Eur. Wind Energy Conf. EWEC*, p 1070, Nice, 1999.
8. H.G. Beyer, J. Luther, and R. Steinberger-Willms. Fluctuations in the combined power output from geographically distributed grid coupled wind energy conversion systems—an analysis in the frequency domain. *Wind Engg.*, **14**:179, 1990.
9. H.G. Beyer, T. Pahlke, W. Schmidt, H.P. Waldl, and U. de Witt. Wake effects in a linear wind farm. *J. Wind Engg. Ind. Aerodyn.*, **51**:303–318, 1994.
10. A.K. Blackadar and H. Tennekes. Asymptotic similarity in neutral barotropic planetary boundary layers. In *Flux of Heat and Momentum in the Planetary Boundary Layer of the Atmosphere*, Final Report, Contract AFCRL-65-531, pp. 1–22, Pennsylvania State University, Meteorological Department, 1968.
11. E. Bossanyi. Short-term wind prediction using Kalman filters. *Wind Engg.*, **9**:1–8, 1985.
12. I.N. Bronstein, K.A. Semendjajew, G. Musiol, and H. Mühlig. *Taschenbuch der Mathematik*. Verlag Harri Deutsch, Thun, Frankfurt am Main, 1993.
13. D.S. Broomhead and G.P. King. chapter On the qualitative analysis of experimental dynamical systems. In Nonlinear Phenomena and Chaos, p. 113. Adam Hilger, Bristol, 1986.

14. J.A. Businger. Transfer of heat and momentum in the atmospheric boundary layer. In *Proc. Arctic Heat Budget Atmos. Circu., RAND Corp.*, pp. 305–332, Santa Monica, CA, 1966.
15. G. Dany. *Kraftwerksreserve in elektrischen Verbundsystemen mit hohem Windenergieanteil*. Ph.D. thesis, RWTH Aachen, 2001.
16. H.A.R deBruin, R.J. Ronda, and B.J.H. van de Wiel. Approximate solutions for the Obukhov length and the surface fluxes in terms of bulk Richardson numbers. *Boundary-Layer Meteorol.*, **95**:145–157, 2000.
17. G. Doms and U. Schättler. *The Non-Hydrostatic Limited-Area Model (Lokalmodell) of DWD, Part I: Scientific Documentation*. Deutscher Wetterdienst (DWD), Offenbach, 1999.
18. Lokalmodell (Deutscher Wetterdienst (DWD)). `http://www.dwd.de`.
19. A.J. Dyer. A review of flux-profile relations. *Boundary-Layer Meteorol.*, **7**:363–372, 1974.
20. B. Ernst. *Entwicklung eines Windleistungsprognosemodells zur Verbesserung der Kraftwerkseinsatzplanung*. Ph.D. thesis, Institut für Solare Energieversorgungstechnik (ISET), Kassel, 2003.
http://opus.uni-kassel.de/opus/volltexte/2003/82/pdf/dis2928_16.pdf.
21. B. Ernst, K. Rohrig, P. Schorn, and H. Regber. Managing 3000 MW wind power in a transmission system operation centre. In *Proc. Eur. Wind Energy Conf. EWEC*, p. 890, Copenhagen, 2001.
22. J. Eichborn et al. A three-dimensional viscous topography mesoscale model. *Contrib. Atmos Phys.*, **70**(4), 1997.
23. B. Everitt. *Cluster Analysis*. Heinemann Educational Books, London, 1974.
24. E. Feitosa, A. Pereira, and D. Veleda. Brazilian wind atlas project. In *Proc. Eur. Wind Energy Conf. EWEC*, pp. 850–853, Copenhagen, 2001.
25. J.H. Ferziger and M. Peric. *Computational Methods for Fluid Dynamics*. Springer, Berlin, 1996.
26. C.A.J. Fletcher. *Computational Techniques for Fluid Dynamics*. Springer, Berlin, 1987.
27. U. Focken. *Leistungsvorhersage räumlich verteilter Windkraftanlagen unter besonderer Berücksichtigung der thermischen Schichtung der Atmosphäre*, vol. 503 of *Fortschritt-Berichte VDI Reihe 6*. VDI Verlag, Düsseldorf, 2003.
28. U. Focken and D. Heinemann. Influence of thermal stratification on wind profiles for heights up to 130 m. In *Proc. Eur. Wind Energy Conf. EWEC*, Madrid, 2003.
29. U. Focken, D. Heinemann, and H.-P. Waldl. Wind assessment in complex terrain with the numerical model Aiolos—implementation of the influence of roughness changes and stability. In *Proc. Eur. Wind Energy Conf. EWEC*, Nice, 1999.
30. U. Focken, M. Lange, and H.P. Waldl. Previento—a wind power prediction system with innovative upscaling algorithm. In *Proc. Eur. Wind Energy Conf. EWEC*, p. 826, Copenhagen, 2001.
31. U. Focken, M. Lange, and H.P. Waldl. Previento—regional wind power prediction with risk control. In *Proc. Global Wind Power Conf.*, Paris, 2002.
32. European Centre for Medium Range Weather Forecast (ECMWF). `http://www.ecmwf.int`.

33. E. Georgieva, E. Canepa, and A. Mazzino. Winds release 4.2 user's guide, electronic version. Technical report, Department of Physics, University of Genova (Genova, Italy) and EnviroComp Institute (Fremont, California), 2002.
34. F.-W. Gerstengarbe and P.C. Werner. *Katalog der Grosswetterlagen Europas (1881–1998) nach Hess und Brezowsky*. Potsdam-Institut für Klimafolgenforschung, Potsdam, Offenbach a.M., 1999.
35. G. Giebel. *On the Benefits of Distributed Generation of Wind Energy in Europe*, vol. 444 of *Fortschritt-Berichte VDI Reihe 6*. VDI Verlag, Düsseldorf, 2001.
36. G. Giebel, L. Landberg, G. Kariniotakis, and R. Brownsword. State-of-the-art on methods and software tools for short-term prediction of wind energy production. In *Proc. Eur. Wind Energy Conf. EWEC*, Madrid, 2003.
37. P. Grassberger. Do climatic attractors exist? *Nature*, **323**:609–611, 1986.
38. G.A. Grell, J. Dudhia, and D.R. Stauffer. A description of the fifth-generation Penn State/NCAR mesoscale model (MM5). Technical Report NCAR/TN-398 + STR, National Center for Atmospheric Research (NCAR), 1995.
39. G. Gross, T. Frey, and C. Land. Die Anwendung eines meteorologischen simulations—modelles zur Berechnung der lokalen Windverhältnisse in komplexem Gelände. In *Proc. German Wind Energy Conf. DEWEK*, Wilhelmshaven, 2002.
40. G. Gross and F. Wippermann. Channeling and countercurrent in the Upper-Rhine valley: numerical simaulation. *J. Clim. Appl. Meteorol.*, **26**:1293–1304, 1987.
41. Concentration Heat and Momentum Ltd. `http://www.cham.co.uk`.
42. B.B. Hicks. Wind profile relationship from the "Wangara" experiments. *Q. J. R. Meteorol. Soc.*, **102**:535–551, 1976.
43. U. Högström. Non-dimensional wind and temperature profiles in the atmospheric surface layer: a re-evaluation. *Boundary-Layer Meteorol.*, **42**:55–78, 1988.
44. A.A.M. Holtslag. Estimates of diabatic wind speed profiles from near-surface weather observations. *Boundary-Layer Meteorol.*, **29**:225–250, 1984.
45. H. Holttinen, T.S. Nielsen, and G. Giebel. Wind energy in the liberalized market—forecast errors in day-ahead market compared to a more flexible market mechnism. In *Proc. IEA Joint Action Symposium on Wind Forecasting Techniques*, pp. 171–182, Norrköpping, 2002.
46. D. Hou, E. Kalnay, and K.K. Droegemeier. Objective verification of the SAMEX '98 ensemble forecast. *Mon. Weather Rev.*, **129**:73–91, 2001.
47. P.S. Jackson and J.C.R. Hunt. Turbulent flow over a low hill. *Q. J. R. Meteorol. Soc.*, **101**:929–955, 1975.
48. N. Jensen. A note on wind generator interaction. Technical Report Risø-M-2411, Risø National Laboratory, 1983.
49. I.T. Jolliffe and D.B. Stephenson, eds. Continuous variables. In Forecast Verification, pp. 97–119. Wiley, Chichester, England, 2003.
50. J.U. Jørgensen, C. Moehrlen, B. O'Gallachoir, K. Sattler, and E. McKeogh. HIRPOM: description of an operational numerical wind power prediction model for large scale integration of on- and offshore wind power in Denmark. In *Proc. Global Wind Power Conf.*, Paris, 2002.
51. L.S. Kalkstein, G. Tan, and J.A. Skindlov. An evaluation of three clustering procedures for use in synoptic climatological classification. *J. Clim. Appl. Meteorol.*, **26**:717, 1987.

52. H. Kantz and T. Schreiber. *Nonlinear Time Series Analysis*. Cambridge University Press, Cambridge, 1997.
53. H. Kapitza and D.P. Eppel. The non-hydrostatic mesoscale model GESIMA. Part I: Dynamical equations and tests. *Beitr. Phys. Atmos.*, **65**:129–146, 1992.
54. G. Kariniotakis and P. Pinson. Evaluation of the More-Care wind power prediction platform. Performance of the fuzzy logic based models. In *Proc. Eur. Wind Energy Conf. EWEC*, Madrid, 2003.
55. I. Katic, J. Høstrup, and N. Jensen. A simple model for cluster efficiency. In *Proc. Eur. Wind Energy Conf. EWEC*, Rome, 1986.
56. HIRLAM (Dutch Weather Service (KNMI)). `http://hirlam.knmi.nl`.
57. D.P. Lalas and C.F. Ratto, eds. *Modelling of Atmospheric Flow Fields*. World Scientific, Singapore, 1996.
58. L. Landberg. Short-term prediction of local wind conditions. Technical Report Risø-R-702(EN), Risø National Laboratory, 1994.
59. L. Landberg. Short-term prediction of the power production from wind farms. *J. Wind Engg. Ind. Aerodyn.*, **80**:207, 1999.
60. L. Landberg, G. Giebel, L. Myllerup, J. Badger, H. Madsen, and T.S. Nielsen. Poor-man's ensemble forecasting for error estimation. In *Proc. Conf. WindPower 2002 (AWEA)*, Portland, 2002.
61. B. Lange. *Modelling the Marine Boundary Layer for Offshore Wind Power Utilisation*, vol. 491 of *Fortschritt-Berichte VDI Reihe 6*. VDI Verlag, Düsseldorf, 2003.
62. B. Lange, H.-P. Waldl, R. Barthelmie, A.G. Guerrero, and D. Heinemann. Modelling of offshore wind turbine wakes with the wind farm program FLaP. *Wind Energy*, **6**:87–104, 2003.
63. M. Lange and D. Heinemann. Accuracy of short term wind power predictions depending on meteorological conditions. In *Proc. Global Wind Power Conf.*, Paris, 2002.
64. W. Leonhard and K. Müller. Balancing fluctuating wind energy with fossil power stations—where are the limits? *Electra (CIGRES)*, **204**:14–18, 2002.
65. Ljung. *System Identification, Theory for the User*. Prentice-Hall, Englewood Cliffs, NJ, 1987.
66. E.N. Lorenz. Deterministic nonperiodic flow. *J. Atmos. Sci.*, **20**:130–141, 1963.
67. H. Madsen, ed. *Models and Methods for Predicting Wind Power*. Elsam, pasas Minggu(pejatan), Jakarta selatan, 1996.
68. I. Marti. Wind forecasting activities. In *Proc. IEA Joint Action Symp. Wind Forecast. Tech.*, pp. 11–19, Norrköpping, 2002.
69. I. Marti, D. Cabezon, J. Villanueva, M.J. Sanisidro, Y. Loureiro, E. Cantero, and J. Sanz. LocalPred and RegioPred. Advanced tools for wind energy prediction in complex terrain. In *Proc. Eur. Wind Energy Conf. EWEC*, Madrid, 2003.
70. I. Marti, T.S. Nielsen, H. Madsen, J. Navarro, A. Roldan, D. Cabezon, and C.G. Barquerro. Prediction models in complex terrain. In *Proc. Eur. Wind Energy Conf. EWEC*, pp. 875–878, Copenhagen, 2001.
71. H.-T. Mengelkamp, H. Kapitza, and U. Pflueger. Statistical-dynamical downscaling of wind climatologies. *J. Wind Engg. Ind. Aerodyn.*, **67/68**:449–457, 1997.
72. F. Mesinger. Dynamics of limited-area models: formulation and numerical methods. *Meteorol. Atmos. Phys.*, **63**:3–14, 1997.

73. ALADIN (French Weather Service (MeteoFrance)). `http://www.cnrm.meteo.fr`.
74. Unified Model (United Kingdom MetOffice). `http://www.met-office.gov.uk`.
75. C. Millais and S. Teske. *Wind Force 12—A Blueprint to Achieve 12% of the World's Electricity by 2020*. European Wind Energy Association (EWEA) and Greenpeace, Brussels and Hamburg, 2003.
76. A.S. Monin and A.M. Obukhov. Basic laws of turbulent mixing in the surface layer of the atmosphere. *Tr. Geofiz. Inst. Akad. Nauk SSSR*, **24**(151):163–187, 1954.
77. K. Mönnich. *Vorhersage der Leistungsabgabe netzeinspeisender Windkraftanlagen zur Unterstützung der Kraftwerkseinsatzplanung*. Ph.D. thesis, University of Oldenburg, 2000.
78. P.E. Morthorst. Wind power and the conditions at a liberalized power market. *Wind Energy*, **6**:297–308, 2003.
79. S. Müller, D. Wüstenberg, and T. Foken. Untersuchungen zu einem Windleistungsprognosesystem mittels künstlicher Neuronaler Netze auf Grundlage der Daten des Messnetzes Meckelnburg-Vorpommerns. In *Proc. German Wind Energy Conf. DEWEK*, pp. 277–280, Wilhelmshaven, 2000.
80. C. Nicholis and G. Nicholis. Is there a climatic attractor? *Nature*, **311**:529–532, 1984.
81. T.S. Nielsen, A. Joensen, H. Madsen, L. Landberg, and G. Giebel. A new reference for predicting wind power. *Wind Energy*, **1**:29–34, 1998.
82. T.S. Nielsen, H. Madsen, and J. Tøfting. Experiences with statistical methods for wind power prediction. In *Proc. Eur. Wind Energy Conf. EWEC*, pp. 1066–1069, Nice, 1999.
83. T.S. Nielsen, H. Madsen, and J. Tøfting. WPPT, a tool for on-line wind power prediction. In: *Wind Forecasting Techniques—Technical report from the International Energy Agency*, pp. 93–115, 2000.
84. T.S. Nielsen, H.A. Nielsen, and H. Madsen. Using meteorological forecasts in on-line predictions of wind power. Technical report, ELTRA, Fredericia, Denmark, 1999.
85. T.S. Nielsen, H.A. Nielsen, and H. Madsen. Prediction of wind power using time-varying coefficient-functions. In *Proc. XV IFAC World Congr.*, Barcelona, 2002.
86. I. Orlanski. A rational subdivision of scales for atmospheric processes. *Bull. Am. Meteorol. Soc.*, **56**:529–530, 1975.
87. T.N. Palmer. Predicting uncertainty in forecasts of weather and climate. *Rep. Prog. Phys.*, **63**:71–116, 2000.
88. H.A. Panofsky. Determination of stress from wind and temperature measurements. *Q. J. R. Meteorol. Soc.*, **89**:85–94, 1963.
89. R.A. Pielke. *Mesoscale Meteorological Modeling*, 2nd ed., International Geophysics Series Vol. 78. Academic Press, San Diego, 2002.
90. R.A. Pielke and M.E. Nicholls. Use of meteorological models in computational wind engeneering. *J. Wind Engg. Ind. Aerodyn.*, **67/68**:363–372, 1997.
91. P. Pinson and G. Kariniotakis. On-line assessment of prediction risk for wind power production forecasts. In *Proc. Eur. Wind Energy Conf. EWEC*, Madrid, 2003.
92. P. Pinson and G. Kariniotakis. On-line adaptation of confidence intervals based on weather stability for wind power forecasting. In *Proc. Global WindPower*, Chicago, 2004.

93. P. Pinson, N. Siebert, and G. Kariniotakis. Forecasting of regional wind generation by a dynamic fuzzy-neural networks based upscaling approach. In *Proc. Eur. Wind Energy Conf. EWEC*, Madrid, 2003.
94. H. Poll, H.-P. Flicke, B. Stern, and H.-J. Haubrich. Kraftwerkseinsatzoptimierung unter Einbeziehung von Strom- und Brennstoffmarkt. *Energ.wirtsch. Tagesfr.*, **1/2**:14–17, 2002.
95. C.A. Poulsen. The mathematical representation of wind speed and temperature profiles in unstable atmospheric surface layer. *J. Appl. Meteorol.*, **9**:857–861, 1970.
96. L. Prandtl. Bericht über Untersuchungen zur ausgebildeten Turbulenz. *Z. Angew. Math. Phys.*, **5**:136, 1925.
97. K. Rohrig. *Rechenmodelle und Informationssysteme zur Integration groer Windleistungen in die elektrische Energieversorgung*. Ph.D. thesis, Institut für Solare Energieversorgungstechnik (ISET), Kassel, 2003. http://opus.uni-kassel.de/opus/volltexte/2004/92/pdf/dis3052_16.pdf.
98. K. Rohrig, B. Ernst, M. Hoppe-Kilpper, and F. Schlögl. Online-monitoring and prediction of wind power in German transmission system operation centres. In *Proc. World Wind Energy Conference (WWEC)*, Capetown, South Africa, 2003.
99. B.H. Saas, N.W. Nielsen, J.U. Jørgensen, B. Armstrup, and M. Kmit. The operational DMI-HIRLAM system. Technical Report 00-26, Danish Meteorological Institute (DMI), Copenhagen, 2000.
100. R. Schrodin. *Quarterly Report of the Operational NWP-Models of the Deutscher Wetterdienst*, vol. 16. Deutscher Wetterdienst, Offenbach a.M., 1998.
101. R. Schrodin. *Quarterly Report of the Operational NWP-Models of the Deutscher Wetterdienst*, vol. 20. Deutscher Wetterdienst, Offenbach a.M., 1999.
102. A.M. Sempreviva, I. Troen, and A. Lavagnini. Modelling of wind power potential over Sardinia. In *Proc. Eur. Wind Energy Conf. EWEC*, 1986.
103. M. Shahgedanova, T.P. Burt, and T.D. Davies. Synoptic climatology of air pollution in Moscow. *Theor. Appl. Climatol.*, **61**:85, 1998.
104. T. Sperling, R. Hänsch, W. Brücher, and M. Kerschgens. Einsatz eines mesoskaligen Strömungsmodells zur Berechnung von flächendeckenden Windenergiepotentialen im gegliederten Gelände. In *Proceedings of the German Wind Energy Conference DEWEK*, pp. 93–96, Wilhelmshaven, 1996.
105. M. Strack. *Analyse und Extrapolation des Windprofils am 130-Meter-Messmast des DEWI*. Studienarbeit, University of Oldenburg, 1996.
106. M. Strack. *Untersuchung zur Anwendung des mesoskaligen Modells GESIMA zur Windpotentialbestimmung in komplexem Gelände*. Diplomarbeit, University of Oldenburg, 1998.
107. R.B. Stull. *An Introduction to Boundary Layer Meteorology*. Kluwer, Dordrecht, 1994.
108. L. Takacs. A two-step scheme for the advection equation with minimized dissipation and dispersion errors. *Mon. Weather Rev.*, **113**:1050–1065, 1985.
109. C. Tauber. Energie- und volkswirtschaftliche Aspekte der Windenergienutzung in Deutschland—Sichtweise von E.ON Kraftwerke. *Energ.wirtsch. Tagesfr.*, **12**:818–823, 2002.
110. R.B. Thomas. *The Old Farmer's Almanac*. Yankee, Dublin, USA, 2003.

111. R.M. Tracy, G.T. Phillips, and P.C. Patniak. Developing a site selection methodology for wind energy conversion systems. Technical Report DOE/ET/20280-3 NTIS, NTIS, 1978.
112. D.J. Tritton. *Physical Fluid Dynamics*. Clarendon Press, Oxford, 1988.
113. I. Troen. The Wasp code. *In Modelling of Atmospheric Flow Fields*, pp. 435–444. World Scientific, Singapore, 1996.
114. I. Troen. On diagnostic wind field models. *In Modelling of Atmospheric Flow Fields*, pp. 503–512. World Scientific, Singapore, 1996.
115. I. Troen and E. L. Petersen. *European Wind Atlas*. Risø National Laboratory, Risø, 1989.
116. H. v.Storch and F.W. Zwiers. *Statistical Analysis in Climate Research*. Cambridge University Press, Cambridge, 1999.
117. H.P. Waldl. *Flap—Wind Farm Layout Program*. University of Oldenburg, Oldenburg, 1996.
118. H.P. Waldl. *Modellierung der Leistungsabgabe von Windparks und Optimierung der Aufstellungsgeometrie*. Ph.D. thesis, University of Oldenburg, 1997.
119. E.K. Webb. Profile Relationship: the log-linear range and extension to strong stability. *Q. J. R. Meteorol. Soc.*, **96**:67–90, 1970.
120. H.R.A. Wessels. Distortion of the wind field by the Cabouw meteorological tower. Technical Report 83-15, Royal Netherlands Meteorological Institute (KNMI), deBilt, 1983.
121. T.H. Wonnacott and R.J. Wonnacott. *Introductory Statistics*. Wiley, New York, 1977.
122. N. Wood. Wind flow over complex terrain: a historical perspective and the prospect for large-eddy modelling. *Boundary-Layer Meteorol.*, **96**:11–32, 1999.
123. A. Yamaguchi and T.A. Ishihara. Dynamical statistical downscaling procedure for wind climate assessment. In *Proc. Eur. Wind Energy Conf. EWEC*, Madrid, 2003.
124. B. Yarnal. *Synoptic Climatology in Environmental analysis—A Primer*. Belhaven Press, London, 1993.
125. A. Zell. *Simulation Neuronaler Netze*. Addison-Wesley, Bonn, 1996.

Index

adaptive estimation 11
adaptive statistical methods 20
adaptivity 13
adiabatic expansion 33
advection equation 70
air pollution 138
Aladin 10
amplitude error 61, 63, 69, 70, 80
arbitrary location 56
Arpege 10
atmospheric boundary layer 2, 8, 19
atmospheric pressure 25, 33, 60, 138, 166, 167
atmospheric stratification 32, 78, 89, 138
auto-correlation 66, 88
automatic classification 135

balance 4, 5, 24, 36, 163, 177, 179
balancing power 5
Bessel function 49, 50
Bessel series 50
between-cluster variance 142
boundary conditions 4, 19, 49
boundary layer VII, 2, 8, 17, 19–21, 23, 24, 26, 29, 31, 39, 41, 42, 47, 52, 78, 91, 101, 103, 180, 203
bulk gradient 103
buoyancy 19, 31, 32, 34, 35, 38, 44, 45, 78, 92
BZ–model 48

Cabouw 57, 92, 94
CENER 20
certified power curve 16, 22, 131, 133, 180
channelling effects 47, 165, 166
chaotic properties 136
classification 92, 135–138, 140, 142, 143, 148, 149, 154, 155, 157, 158, 160, 161, 164–167, 180
climatological mean 43, 68, 88
climatology 3, 135, 137, 148, 166, 179, 180
cluster analysis 20, 138, 140, 144, 148, 149, 166
cluster distance 141, 149, 150
complete linkage method 141
complex terrain 15, 19, 47, 48, 74, 78, 79, 82, 86, 89, 167, 179
computational grid 8, 55, 56
conditional pdf 115–123, 125–127, 132
confidence interval 62, 64, 66, 68, 70, 75, 79, 125, 126, 158, 159, 187, 190–192, 194
conservation of mass 19, 23, 26
conservation of momentum 25
continuity equation 26
conventional power plants 1, 3, 4, 6, 7
cooling 33, 44
Coriolis force 24, 25, 36, 37
cost-efficient energy supply 4

coupling 24, 25, 36–38, 43, 44, 78, 92, 94, 109
covariance matrix 139
cross-correlation 61, 69, 70, 72, 73, 80, 86, 87, 90, 97–101, 105, 107, 111, 172, 173, 175, 177, 178, 183
cup anemometer 55
cut-in speed 68, 117
cylindrical polar coordinates 49

daily forecast error 163, 167
data matrix 139, 140, 144, 148
day ahead 6
daytime 44, 78, 82, 107, 109, 110, 113, 114
decomposition 27, 69, 70, 72, 75, 80, 81, 86, 89
degree of freedom 190
Denmark V, 5, 10, 11
detailed roughness description 2
diagnostic level 56
Diagnostic models 21
direction dependent 21, 39, 43, 108
dispersion 70, 72, 73, 80, 81, 87–90, 184
distribution of distance 175
diurnal variation 78
diurnal cycle 32, 44, 109, 111, 114
Diurnal variation 78, 89
diurnal variation 15, 78, 80, 82, 89, 110, 111, 114, 145, 147
dry adiabatic temperature gradient 33
dry gas constant 60
dynamic viscosity 25

ECMWF 136
economic value 6
effective non-linearity factor 81, 87
eigenmode 138–140, 148
eigenvalue 139, 144, 145, 148
eigenvector 137, 139, 140
Ekman layer 23, 31
electricity V, VI, 1, 3–7, 169, 179
ELSAM 11
ELTRA 11
empirical pdf 126
energy market 4
energy supplier 1
ensemble of wind farms 169, 177
equations of motion 2, 8, 18, 20, 23, 26, 37, 47, 49, 55, 56, 136
equidistant bin 120
error characteristic 68
error source 69, 128, 133, 180
estimation procedure 11
European Weather Service 9
European Wind Atlas 21, 37, 41, 48
evaluation 3, 68, 144
EWEA 1
extreme situation 163

F test 143, 158, 163, 189
Fitnah 19, 47
flat terrain 74, 78, 79, 82, 83, 86, 89, 91, 108, 111, 177
flow distortion 58, 66, 165, 166
flow effect 19, 25
fluctuation V, 1, 3, 6, 26–30, 32, 34, 38, 56, 70, 75, 79, 86, 90, 169, 189
forecast accuracy 62, 114, 121
forecast error 2, 3, 6, 21, 39, 61, 64, 67, 72, 80, 81, 89, 90, 115, 126, 130, 135, 138, 142, 144, 148, 150, 158–161, 163–167, 169, 170, 172, 173, 176
forecast situation 115, 116, 132, 180
forecasting system 2, 20, 61
forgetting factor 13–15
Fourier series 50
friction velocity 31, 32, 36, 37, 42, 47
frontal system 138

Gaussian 15, 61, 64, 65, 67, 90, 115–122, 127, 132, 133, 187, 188
geostrophic drag law 37, 41, 43
geostrophic wind 23–25, 36, 37, 41, 94
Gesima 47
global model 9, 56
Globalmodell 10
gradient method 45
gravitational constant 60
gravity 25, 36

Greenpeace 1
grid cell 8, 41, 55, 79, 89
grid domain 19
grid operator 1
Grosswetterlage 137
ground level 44, 45, 49, 60

heat capacity 32, 34, 78, 180
heat flux 32, 34, 43–45, 47
hierarchical 140
high grid penetration 4, 5
high-pressure 3, 135, 136, 147, 152–161, 163, 165, 166
hill 20, 21, 47, 79, 89, 144, 165, 166
HIRLAM 15, 20, 21, 64
horizontal momentum 24, 29–31, 37, 44, 93
hub height 3, 8, 17, 39, 82, 89, 91, 95, 107, 109, 111, 113, 114, 120, 179
hydrostatic 19, 105

IBL 41–43
individual correction 98
individual site 61, 74, 86, 172
initial condition 9, 48, 88, 136
installed wind power capacity 1
integration of wind energy 1
internal boundary layer 41, 42, 180
inverse distance 56
inversion 44, 82
irrotational 49
ISET VIII, 15, 20
isobar 24, 36

Jackson and Hunt 48
joint distribution 116, 117

k-step prediction 14
kinematic viscosity 28, 29

Laplace equation 49
large synoptic pressure system 36
large-scale weather map 135, 144, 152, 154, 166
least-squares estimation 11
limit distribution 175
linear regression 11, 39, 71, 72, 81, 87, 90, 121, 124, 132
linear transformation 61, 70–73, 81, 90, 98
linkage method 141, 149, 203
load profile 4
local effect 89, 165–167
local refinement 2, 11, 20, 21, 82, 90
local thermal effect 105
local wind condition 23, 87, 181
LocalPred 15
logarithmic profile 32, 35, 48, 51, 91, 96, 98, 100, 101, 113
logarithmic wind profile 2, 3, 32, 40, 41, 43, 56, 79, 81, 113
Lokalmodell 10
look-up table 20, 50
lower atmosphere 3, 25, 29, 38, 43, 91
lower boundary layer 78, 91, 101, 103

main run 56
manual method 137
marine boundary layer 180
marine boundary layer 52
MASS 6 20
mass-consistent model 19, 20
matching relation 43
MATLAB 144
meso-scale 15, 19, 20, 43, 47, 79, 89, 179, 181
meteorological mast 57
micro-scale 18, 19
mixing length 28–30, 38
model level 56
model output statistic 70
molecular friction 25, 36
molecular viscosity 25, 28, 38
momentum transport 31, 49, 50, 78, 94, 109
Monin–Obukhov theory 2, 3, 23, 32, 57, 95, 98, 101, 113
MORE-CARE 17
multi-dimensional power curve 16

Navier–Stokes 28

Navier–Stokes equation 19, 25–27
negative load 4, 5
nesting 19, 20
neutral stratification 36, 96
Newtonian fluid 25, 28
nighttime 44, 78, 82, 109, 110
noise sensitivity 13
non hydrostatic 19, 105
non-Gaussian 67
non-linear power curve 90
non-parametrical Lilliefors test 64
nonlinear power curve 3, 61, 68, 115, 180
normal distribution 64–66, 121–123, 187
numerical grid 8, 50, 181
numerical integration 55
numerical model 18, 20, 26, 47, 48
NWP 2, 7–9, 11, 17, 19–21, 23, 39, 41, 43, 44, 56, 57, 64, 78, 79, 82, 87–90, 105, 107, 109, 111, 113, 120, 135, 136, 165, 177, 181

offset 70
offshore 52, 78, 132, 179, 180
on-site condition 8, 17, 41, 74, 89, 165
online monitoring 15
online operation 13
orographic effect 2, 21, 49, 108, 144, 165
orography 21, 39, 74, 79, 89, 165, 166
orthogonal basis 50, 148
orthonormal basis 139
overestimation 62, 79, 82, 86, 89, 97, 109

parameterisation 15
persistence 21, 68, 69, 87, 88
phase error 61–63, 69, 70, 79–81, 85, 87, 89, 143
phase space 139–141, 148, 149
physical approach VII, 15, 21, 39
physical parameterisation 15
physical prediction system 2, 17, 22
physical system 2, 8, 10, 21, 22
point prediction 56, 66
pointwise error 62, 64, 172, 184
pointwise prediction error 62, 172
poor man's ensemble forecast 136
post-processing 70
potential flow 49, 50
potential temperature 33, 36, 92
Prandtl layer 31, 37
predictability 105, 113, 165, 167
predictable V, 136
prediction accuracy 3, 6, 22, 61, 68, 70, 73, 74, 81, 106, 107
prediction uncertainty VII, 3, 131, 135, 177
Prediktor 21
pressure gradient 24, 25, 138, 153, 155, 159, 160
principal component analysis 135, 137, 166
probability 62, 64, 68, 115–119, 132, 136, 187, 188, 190

rated power 63, 67, 81, 82, 109, 117, 127, 128, 131
reconstruction procedure 119, 126, 127
reference model 88
reference system 68, 69, 87, 88
regional prediction 170–173
regional smoothing effect 3, 177
renewable energy VI, 1, 6
representative site 16, 21, 40, 179
Reynolds averaging 26
Richardson number 33, 34, 38, 45, 46
risk index 136
rotational speed of the earth 25, 37
roughness 2, 11, 19, 21, 22, 31, 37, 39–43, 58, 79, 89, 95, 96, 101, 113, 144, 180, 204
roughness change 50
roughness length 50, 101

saturation level 173, 175, 177
SCADA 17
scheduling scheme 7
Scheffe's multiple comparison 143, 190–193
sde 70, 75, 79, 80, 86, 107, 113, 121, 126, 129–131, 184
sea breeze 8

self-calibrating 11, 15
sensible heat flux 45
sensor carrying boom 57
shadowing effect 2, 21, 39, 50, 53, 57, 107
short-term prediction 2, 7, 17, 43
signal to noise ratio 174
significance level 64, 66, 188, 190
similarity 69
single site 3, 169–173, 176, 177
singular perturbation method 37
situation dependent assessment 3, 132
skill score 68
small perturbation 48
smoothing effect 3, 40, 169, 170, 177, 179, 180, 206
solar irradiation 19, 44, 78
sonic anemometer 45
spatial average 55, 56
spatial correlation function 169
spatial refinement 17, 41
spatial resolution 8–10, 18–20, 23, 39, 56
speed-up effect 47, 165, 166
spot market 6
stability parameter 98, 111
stability classes 91, 92, 94
stability correction 56, 101, 103, 105, 107, 109, 111, 113
stability parameter 34, 45, 91, 96, 98, 103, 109
stable high-pressure 3, 153
stable stratification 91–95, 98, 103, 105, 109
standard deviation 64–66, 69, 70, 72, 74, 75, 79–82, 86, 87, 89, 98, 100, 101, 105, 107, 111, 115, 118–122, 125–127, 129, 132, 139, 171, 172, 183, 184
state of the art 179
stationary situation 95, 98
statistical approach 7, 17
statistical behaviour 3, 8, 61, 66, 73, 74
statistical correction 21, 39, 70
statistical distribution 62, 67, 91, 95, 107, 132
statistical error measure 68, 69, 71, 75, 97
statistical significance 143
statistical system 2, 8, 10, 11, 15, 22, 39
statistical test 64, 121, 122, 129, 158, 187
stochastic process 11
sub-grid scale 8
subspace 139, 140
surface layer VI, 23, 31, 35–37, 44, 94, 100, 109, 113
surface pressure 60, 139, 144
surface roughness 21, 22, 31, 39, 40, 42, 43, 89, 144, 180
surface temperature 60
synoptic climatology 3, 135, 137, 166, 180
synoptic scale 56, 60
synoptic weather system 8
systematic forecast error 39

Taylor expansion 118, 129
temperature difference 43, 45, 47, 96, 103–107, 111, 113, 138
temporal evolution 62, 71, 135, 145
temporal mean 69
terrain type 31, 78, 79, 82, 89
test case 3, 74
thermal correction 2, 3, 35, 95–97, 100, 101, 104, 109, 110, 113
thermal effect VII, 19, 25, 26, 31, 32, 38, 100, 105, 205
thermal stratification 2, 3, 15, 21, 23, 32–34, 37–39, 43, 45, 47, 82, 89, 91, 94, 96, 98, 99, 101, 105, 107, 109, 111, 113, 114
thermally corrected profile 97, 98, 101
thermally corrected wind profile 44
thermally induced turbulence 45
time-delay embedding 139
time-varying parameter 13, 15
topographic structure 47
total variance 144, 145
training period 16
transmission system operator 4, 169, 179
TSO 4–6, 11, 15, 39
turbulence VII, 8, 18, 19, 25, 27–29, 31–34, 38, 45, 49, 56, 207
turbulent fluctuation 27, 28, 32

turbulent mixing 25, 38, 43, 44, 92
turbulent momentum flux 31, 38
turbulent momentum transport 31, 49, 50, 109
turbulent motion 27

uncertainty VII, 3, 21, 62, 98, 107, 116, 131, 133, 135, 136, 167, 177, 179, 180, 206
unconditional error distribution 117
uncorrelated source 174
underestimation 62, 79, 86, 89, 97–99
Unified Model 10
University of Oldenburg 2, 21, 39
unstable stratification 93, 103, 105, 109
up-scaling 21, 40, 179
UTC 56, 87, 136, 138, 139, 167

value of wind energy 3
variance of the error 69, 143, 184
velocity gradient 25, 30, 34, 35, 38, 92
verification VI, 73, 127, 171
vertical heat flux at the surface 34
vertical velocity gradient 30, 34
vertical wind profile 23, 25, 31, 32, 37, 41, 91, 101
visual inspection 61, 64, 139
von Karman constant 31, 37

wake effect 51, 52, 180
WAsP 19, 20
weather situation 136, 148, 155
weather classes 135, 137, 140, 144, 150, 152–154, 157, 160, 165, 167
weather condition 3, 135, 142, 148, 160, 166, 177
weather situation 3, 135–138, 150, 151, 154–159, 161, 163–167
Weibull distribution 75, 120
weighting factor 16
wind shear 31, 45
within-cluster variance 142, 143
WMEP measurement data 62, 64
WPMS 15, 16, 20
WPPT 11, 14, 15

Printing: Krips bv, Meppel
Binding: Stürtz, Würzburg